# Concise Textbook of Equine Clinical Practice Book 2

This concise, practical text covers the essential information veterinary students and nurses, new graduates and practitioners need to succeed in equine practice, focussing on reproduction and the foal. Written for an international readership, the book conveys the core information in an easily digestible, precise form with extensive use of bullet-points, lists, diagrams, protocols and extensive illustration (over 200 full colour, high quality photographs).

Part of a five-book series that extracts and updates key information from Munroe's *Equine Surgery, Reproduction and Medicine*, Second Edition, the book distils best practice in a logical straightforward clinical-based approach. It details clinical anatomy, physical clinical examination techniques, diagnostic techniques and normal parameters, emphasising the things regularly available to general practitioners with minimal information of advanced techniques. The clinical information is split into anatomy-based sections.

Ideal for veterinary students and nurses on clinical placements with horses as well as for practitioners needing a quick reference 'on the ground'.

# Concise Textbook of Equine Clinical Practice Book 2

# Reproduction and the Foal

**Tracey Chenier**
**Charles D. Cooke**
**Graham Munroe**
**Victoria Scott**

**Edited By**

**Graham Munroe**

**CRC Press**
Taylor & Francis Group
Boca Raton London New York

CRC Press is an imprint of the
Taylor & Francis Group, an **informa** business

First edition published 2024
by CRC Press
6000 Broken Sound Parkway NW, Suite 300, Boca Raton, FL 33487-2742

and by CRC Press
4 Park Square, Milton Park, Abingdon, Oxon, OX14 4RN

*CRC Press is an imprint of Taylor & Francis Group, LLC*

*Library of Congress Cataloging-in-Publication Data*

Names: Chenier, Tracey, author. | Cooke, Charles D., author. | Munroe, Graham A., author. | Scott, Victoria (Victoria Helen Louise), author.

Title: Concise textbook of equine clinical practice. Book 2, Reproduction and the foal / Tracey Chenier, Charles D. Cooke, Graham Munroe, Victoria Scott ; edited by Graham Munroe.

Other titles: Reproduction and the foal | Equine clinical medicine, surgery and reproduction. 2nd edition.
Description: First edition. | Boca Raton : CRC Press, 2023. | Abridgement of: Equine clinical medicine, surgery and reproduction / edited by Graham Munroe. 2nd edition. 2020. | Includes bibliographical references and index. | Summary: "This concise, practical text covers the essential information veterinary students, new graduates and practitioners need to succeed in Equine practice, focussing on reproduction. Written for an international readership, the book conveys the core information in an easily digestible, precise form with extensive use of bullet-points, lists, diagrams, protocols and extensive illustrations. Part of a four-book series that extracts and updates key information from Munroe's Equine Surgery, Reproduction and Medicine, Second Edition, the book distils best practice in a logical straightforward clinical-based approach. The spiralbound format allows the book to lie open during practice"--

Provided by publisher.
Identifiers: LCCN 2023006706 (print) | LCCN 2023006707 (ebook) | ISBN 9781032479828 (hardback) | ISBN 9781032066189 (paperback) | ISBN 9781003386834 (ebook)
Subjects: MESH: Reproduction--physiology | Horses--physiology | Reproductive Techniques--veterinary | Breeding--methods | Pregnancy Complications--veterinary | Genital Diseases--veterinary
Classification: LCC SF887 (print) | LCC SF887 (ebook) | NLM SF 887 | DDC 636.089/82--dc23/eng/20230525

LC record available at https://lccn.loc.gov/2023006706
LC ebook record available at https://lccn.loc.gov/2023006707

ISBN: 978-1-032-47982-8 (hbk)
ISBN: 978-1-032-06618-9 (pbk)
ISBN: 978-1-003-38683-4 (ebk)

DOI: [10.1201/9781003386834]

Typeset in Sabon
by Evolution Design & Digital Ltd (Kent)

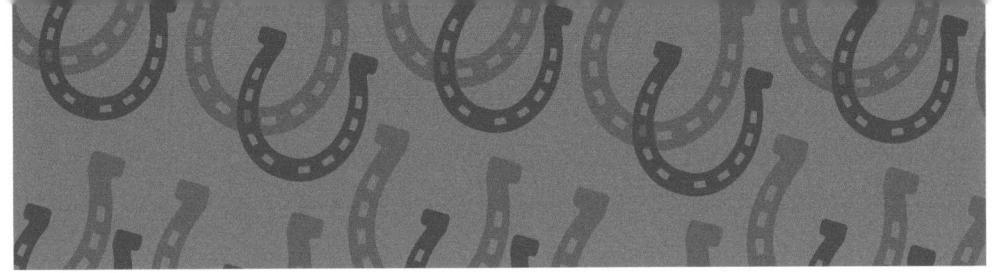

# Table of Contents

# Preface

There is a vast array of clinical equine veterinary information available for the under- and post graduate veterinarian and veterinary nurse to peruse. This is contained in textbooks, both general and specialised, and increasingly online at websites of varying quality and trustworthiness. It is easy for the veterinary student or nurse, recent graduate, and busy general or equine practitioner to become overwhelmed and confused by this diverse range of information. Often what is needed, particularly in the clinical situation, is a distillation of the essential knowledge and best-practice required to treat the horse in the most suitable way. This concise, practical text is designed to provide that essential information needed to understand and treat clinical cases in equine practice.

This book focuses on reproduction in the stallion and mare, plus the result of the reproductive process, namely the foal. It is part of a five-book series, which between them will cover all the areas of equine clinical practice. The information is extracted and updated from *Equine Clinical Medicine, Surgery and Reproduction* (2nd edition) that was published in 2020. It is written for an international readership and is designed to convey the core, best-practice information in an easily digested, quick reference form using bullet-points, lists, tables, flowcharts, diagrams, protocols, and extensive illustrations and photographs.

The mare part of the book is split into sections covering the normal physiology and anatomy, methods of examination, breeding management, pregnancy and complications, parturition, veneral diseases, and specific diseases of the anatomical regions of the female reproductive tract. In the stallion, sections cover the normal anatomy and physiology, methods of examination and assessment, veneral and congenital diseases, and conditions of the specific parts of the reproductive tract. For the foal, sections cover the examination of the foal and its normal parameters, drug therapy, routine care, and specific conditions of the foal, both neonate and older foal. The information is set out in the same logical straightforward clinical-based way. The clinical diseases are all covered using the same format of aetiology/pathophysiology, clinical examination findings, differential diagnosis, diagnostic techniques, management and treatment, and prognosis. The emphasis is on information tailored to general equine clinicians with just enough on advanced techniques to make the practitioner aware of what is available elsewhere.

The intention of this series of books is for them to be used on a day-to-day basis in clinical practice by student and graduate veterinarians and nurses. The small size and spiral binding format allow them to lie open on a surface near to the patient, readily available to the veterinary student or practitioner whilst looking at, or treating, a clinical case.

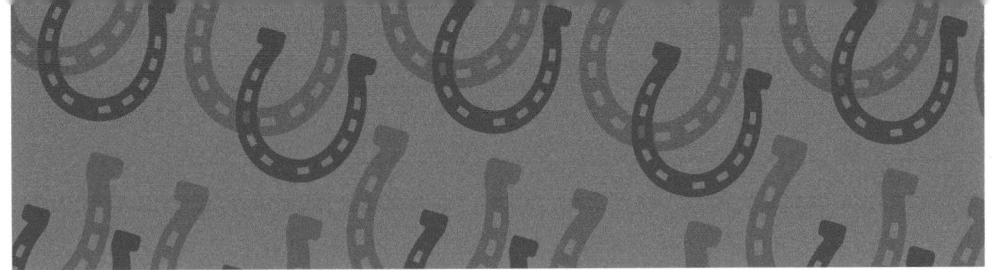

# Author Biography

**Tracey Chenier, DVM, DVSc, DACT**
Tracey Chenier is an Associate Professor of Theriogenology at the Ontario Veterinary College, University of Guelph, in Guelph, Canada. She attained Board Certification with the American College of Theriogenologists in 1996. Her areas of interest and research include stallion reproduction, semen cryopreservation and endometritis in mares.

**Charles D. Cooke BSc (Hons) BVetMed CertEM (Stud Medicine) MRCVS**
Charles D. Cooke qualified from the Royal Veterinary College, University of London in 2000 and works in general equine practice focusing on all aspects of equine stud medicine including natural breeding, artificial insemination and embryo transfer/assisted reproductive techniques in all types of horses and ponies. He has a particular interest in young-stock management and preventative medicine on the stud farm, treatment of the 'sub-fertile' mare, foetal sexing and the management of hormone-related behavioural problems in competition horses. He was awarded the RCVS Certificate in Equine Medicine (Stud Medicine) in 2008 and is an RCVS Advanced Practitioner; he has been an honorary lecturer at the University of Liverpool since 2014 and became a director at Equine Reproductive Services (UK) in 2016.

**Graham Munroe** qualified from the University of Bristol with honours in 1979. He spent 9 years in equine practice in Wendover, Newmarket, Arundel, and Oxfordshire, and a stud season in New Zealand. He gained a certificate in equine orthopaedics and a diploma in equine stud medicine from the RCVS whilst in practice. He joined Glasgow University Veterinary School in 1988 as a lecturer and then moved to Edinburgh Veterinary School as a senior lecturer in large animal surgery from 1994 to 1997. He obtained FRCVS in 1994 and DipECVS in 1997 by examination, and was awarded a PhD in 1994 for a study in neonatal ophthalmology. He has been visiting equine surgeon at the University of Cambridge Veterinary School, University of Bristol Veterinary School and Helsingborg Hospital, Sweden and was team veterinary surgeon for British Driving Teams from 1994 to 2001, the British Dressage Team from 2001 to 2002 and the British Vaulting Team in 2002. He was also FEI veterinary delegate at the Athens 2004 Olympics. He currently works in private referral surgical practice, mainly in orthopaedics. He has published over 60 papers and book chapters.

**Victoria Scott BVetMed MSc DACVIM MRCVS**
Victoria qualified from the Royal Veterinary College, University of London in 2006 and is presently a PhD candidate at the University of Glasgow. She worked at Rossdale and Partners in Newmarket, UK as an intern and equine stud farm veterinarian for 3 years. She spent 3 years in an internal medicine residency at The Ohio State University, USA where she obtained a MSc and board certification in equine internal medicine. She was an equine internal medicine clinician and lecturer at Cambridge University Veterinary School from 2013 to 2019. She has interests in all aspects of foal and internal medicine.

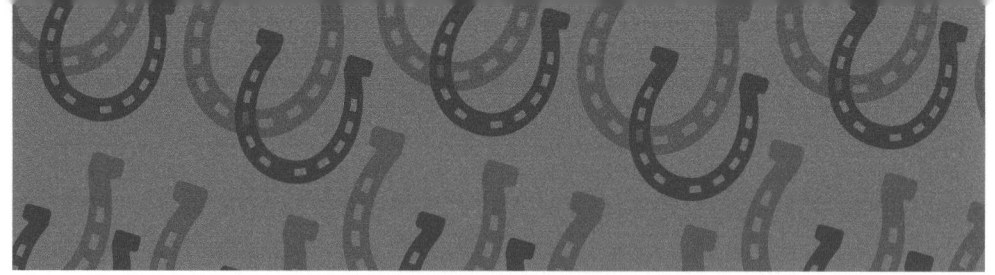

# Table of Abbreviations

| | |
|---|---|
| ACTH | adrenocorticotropic hormone |
| AI | artificial insemination |
| AMH | anti-müllerian hormone |
| AP | alkaline phosphatase |
| AV | artificial vagina |
| BCG | bacillus calmette-guérin |
| BSE | breeding soundness examination |
| BWT | body weight |
| CARS | compensatory anti-inflammatory response syndrome |
| CASA | computer-assisted sperm motion analysis |
| CBC | complete blood count |
| CCP | corpus cavernosum penis |
| CEM | contagious equine metritis |
| CNS | central nervous system |
| CRI | constant rate infusion |
| CTUP | combined thickness of the uterus and placenta |
| DSD | disorders of sexual development |
| DUI | deep uterine insemination |
| ECG | electrocardiogram |
| EED | early embryonic death |
| EHV | equine herpesvirus |
| EIA | equine infectious anaemia |
| EVA | equine viral arteritis |
| FSH | follicle stimulating hormone |
| FTPI | failure of transfer of passive maternally derived immunity |
| GA | general anaesthetic |
| GCT | granulosa cell tumour |
| GDUD | gastroduodenal ulceration disease |
| GI | gastrointestinal |
| GIT | gastrointestinal tract |
| GnRH | gonadotrophin-releasing hormone |
| HAF | haemorrhagic anovulatory follicle |
| HYPP | hyperkalaemic periodic paralysis |
| IgG | immunoglobulin g |
| IgM | immunoglobulin m |
| LH | luteinising hormone |
| NMS | neonatal maladjustment syndrome |
| PAF | persistent anovulatory follicle |
| PCR | polymerase chain reaction |
| PCV | packed cell volume |
| PG | prostaglandin |
| PMIE | persistent mating-induced endometritis |
| PMMA | polymethylmethacrylate |
| PMMN | progressively motile morphologically normal |
| PMN | polymorphonuclear (cell) |

| RBC | red blood cell |
| SAA | serum amyloid a |
| SCC | squamous cell carcinoma |
| SIRS | systemic inflammatory response syndrome |
| TD | testicular degeneration |
| TSW | total scrotal width |
| VC | vulval conformation |
| WBC | white blood cell |

# The Female Reproductive Tract

## Introduction

### The oestrous cycle (Table 1.1)

- mares are seasonally polyoestrous breeders:
  - multiple oestrus periods during a 'breeding season'.
- most mares will cycle during the spring/summer/early autumn (long days).
- most mares enter a winter anoestrous period when they stop cycling:
  - about 20% of mares continue to ovulate throughout the winter months.
- **photoperiod** (day length):
  - most important factor influencing seasonality and ovarian activity.
  - gradual transition periods link summer cyclicity and winter anoestrous periods in most mares.
- during winter (short days) most mares are reproductively silent:
  - Pineal gland releases melatonin during hours of darkness.
  - increased melatonin suppresses release of gonadotrophin releasing hormone (GnRH) from the hypothalamus.
- increasing day length (spring) reduces melatonin pulses leading to:
  - increased pulsatile GnRH release.
  - stimulates luteinising hormone (LH) and follicle stimulating hormone (FSH) release from the pituitary gland.
- ovaries develop multiple small–medium follicles:
  - 'bunches of grapes' on rectal palpation and ultrasound.
  - low levels of oestrogen are produced.
- follicular waves continue to develop and regress until sufficient stimulation for a dominant follicle to emerge that grows:
  - oestrogen levels increase and lead to an LH surge and the first ovulation.

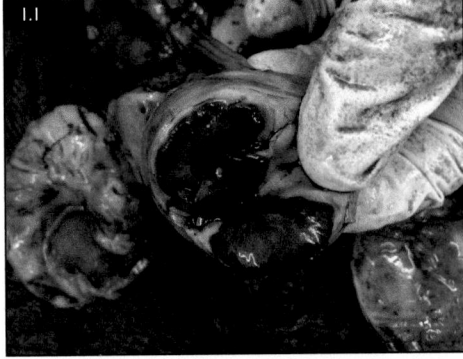

**FIG 1.1** Examination of an ovary at post-mortem clearly demonstrating the CL within the ovary.

  - ovulation is generally about 24 hours before the end of oestrus.
- ovulation initially results in a *corpus haemorrhagicum*:
  - organises, luteinises and contracts into a *corpus luteum* (CL) (Fig. 1.1).
  - progesterone levels increase rapidly causing:
    - uterine oedema to regress, uterine tone to increase and cervix to close.
  - negative feedback on GnRH secretion.
  - FSH continues to stimulate follicular development, but the mare is not 'in-season'.
- prostaglandin F$_2\alpha$ (PGF$_2\alpha$) is released by the endometrium:
  - causes CL regression and a reduction in progesterone.
  - GnRH levels increase, dominant follicle develops, and mare returns to oestrus.
- mare continues to cycle and ovulate during the remainder of the season.
- autumn transition period, the cycle slows in response to environmental factors.
- mare enters winter anoestrous.

DOI: 10.1201/9781003386834-1

**Oestrous cycle:** time period from one ovulation to another:

19–24 days, average 21 days.

**Oestrus:** time when the mare shows 'heat', is receptive to the stallion and stands to be mated:

4–9 days, average 6 days.

dominant hormone – oestrogen.

**Dioestrus:** time between oestrus periods when the mare is not receptive to the stallion:

12–16 days, average 14 days.

dominant hormone – progesterone.

## Monitoring the phases of the oestrous cycle

### Methods of examination

- mares are commonly examined by rectal palpation and ultrasound to determine their stage of the cycle and allow precise management and breeding of the mare.
- **rectal palpation** can assess:
  - ovary size
  - follicle size (approximate)
  - follicle firmness.
  - uterine and cervical tone.
- **ultrasonography** of the reproductive tract gives further detail of:
  - precise follicle size and shape.
  - presence or absence of a CL.
  - follicular abnormalities e.g. haemorrhagic follicular fluid.
  - endometrial oedema.
  - uterine abnormalities e.g. cystic structure, luminal fluid, adhesions.
- **Cervix**
  - manual and visual examination of the cervix must only be performed following thorough cleaning of the vulva with paper towel and water, and then dried.
  - disposable speculum and pen torch (Fig. 1.2) allows full assessment of the vaginal vault and cervix:
    - cervical adhesions and damage
    - vaginal varicose veins.
    - swabs from cervix/uterus for laboratory culture/cytology can be taken.

**TABLE 1.1** Summary of the mare's reproductive findings during the year.

|  | Winter[1] | Spring transition | Summer | Autumn transition |
|---|---|---|---|---|
| Ovary | Small follicles <20 mm | Multiple follicles 20–30 mm<br><br>Grow and regress | Follicular waves and selection of dominant follicle which grows to 35–40 mm, ovulates and forms *corpus haemorrhagicum* then *corpus luteum* | Multiple large follicles that do not ovulate |
| Uterus | Flaccid | Initially flaccid, develops uterine oedema during transition | Oestrus (oestrogen) – low tone, uterine oedema Dioestrus (progesterone) – high tone, no oedema | Uterine oedema becoming flaccid |
| Cervix | Pale, dry partially relaxed | Becoming pink and relaxed | Oestrus – pink, relaxed Dioestrus – pale, closed | Becoming pale |
| Behaviour | Ambivalent | Becoming receptive | Oestrus – receptive to stallion Dioestrus – aggressive to stallion | Receptive becoming ambivalent |
| Hormones | LH low<br><br>FSH fluctuates<br><br>Progesterone < 3.18 nmol/l < 1 ng/ml | LH initially low but increases towards ovulation<br><br>FSH rises early on but falls 15–20d pre-ovulation |  |  |

1 – up to 20% of mares continue to cycle during the winter.

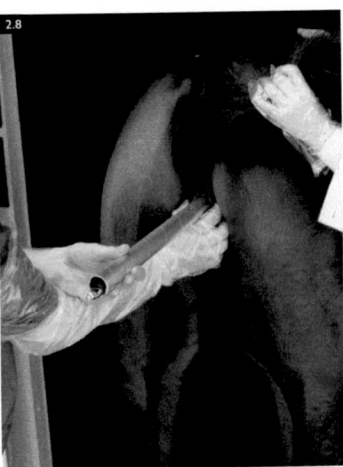

**FIG. 1.2** Introduction of a sterile disposable vaginal speculum to allow visualisation.

- manual cervical examination for relaxation and cervical tear/injury detection.
- **External genitalia**
  - vulval assessment:
    - ◆ previous foaling injury
    - ◆ conformation and risk of pneumovagina.
  - clitoral assessment:
    - ◆ collection of pre-breeding swabs.
- **Endocrinology**
  - blood sample for serum progesterone can be helpful in:
    - ◆ detecting active CLs in the absence of obvious ultrasound features.
    - ◆ mares that are not coming into oestrus as expected.

- ◆ level over >2 ng/l (>6 nmol/l) indicates active luteal tissue.
- **Record keeping**
  - accurate records are essential for:
    - ◆ monitoring and planning of mating/covering the mare.
    - ◆ organising follow up examinations or pregnancy check.
    - ◆ following should be recorded after each examination as appropriate:
      - – follicle size (mm), shape, firmness and follicular fluid appearance.
      - – uterine tone, oedema and fluid.
      - – cervix tone or appearance.
      - – medications given e.g. ovulatory agents, intrauterine treatments etc.
      - – other notes:
        - ◊ management plan for the mare, such as covering/ treatment.
        - ◊ highlight a potential problem.
      - – next examination date.

## Reproductive ultrasound examination

### Ovarian ultrasound

- **Follicle:**
  - hypoechoic, spherical structures (Figs. 1.3–1.6).
  - matures and prepares to ovulate:

**FIG. 1.3** Ultrasonographic image of an ovary of a mare in oestrus showing a large developing follicle.

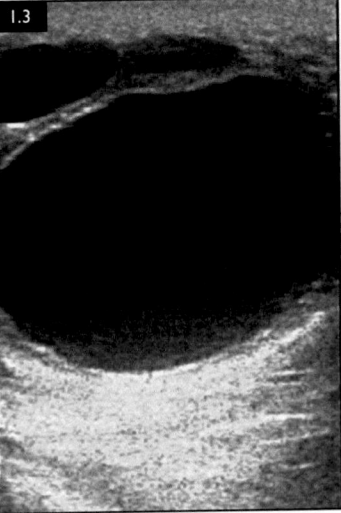

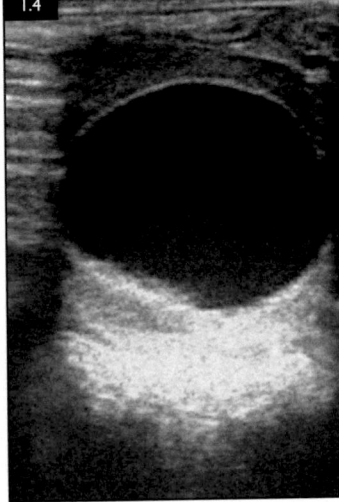

**FIG. 1.4** Ultrasonographic image of an ovarian follicle just prior to ovulation.

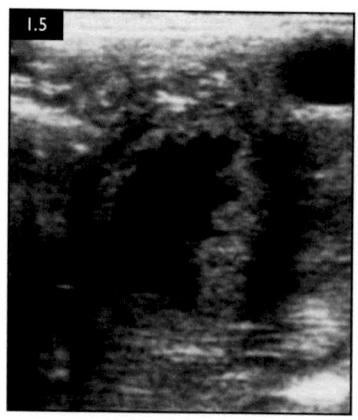

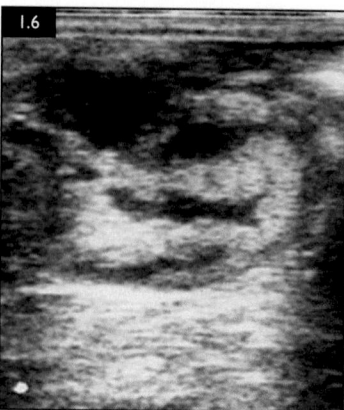

**FIG. 1.5** Ultrasonographic image of an ovarian follicle during ovulation.

**FIG. 1.6** Ultrasonographic image of an ovarian follicle just after ovulation has occurred.

- wall thickens and follicle seen to 'point' towards the ovulation fossa.
  - during ovulation the follicle collapses.
- **Corpus luteum**
  - echogenic structure which can vary in echogenicity and size (Figs. 1.7–1.9).
  - some contain fluid areas (lacuna) that are a normal variation:
    - do not alter progesterone production.
  - Doppler examination allows blood flow to be visualised:
    - related to 'CL health and activity' and progesterone production.

## Uterine ultrasound

- uterus is a tubular 'Y' shaped structure:
  - horns seen in transverse section (circular) and the body in longitudinal section (rectangular) (Fig. 1.10).

- uterine oedema is present when the mare is in oestrus (Fig. 1.11)
  - appears as a 'cart-wheel' or 'cut orange' appearance.
  - excessive oedema can suggest a uterine lymphatic issue or that the mare may have a pronounced post-covering reaction.
- uterine fluid can be seen in the lumen (Fig. 1.11)
  - character of the fluid can vary from hypoechoic to hyperechoic to purulent.

## Identifying when the follicle may ovulate

- mare has a relatively long oestrus period (5–7 days):
  - ovulation occurs 24 hours prior to the end of behavioural oestrus.

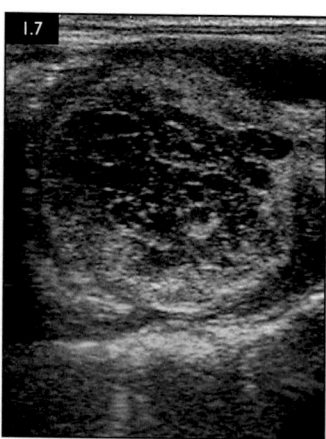

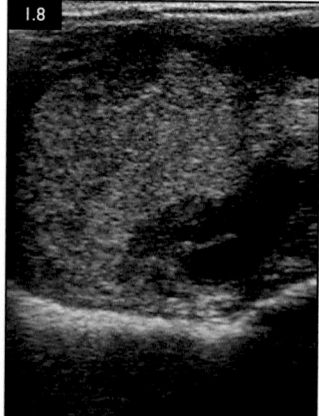

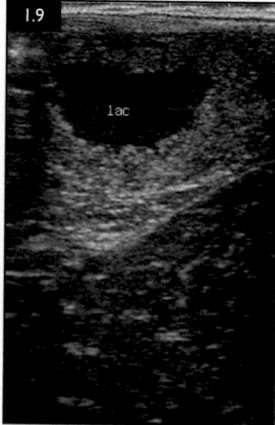

**FIGS. 1.7–1.9** (1.7) Ultrasonographic image of an ovary showing an early developing CL. (1.8) A mature CL showing a more organised echogenic structure compared with the one in 1.7. (1.9) A normal variation on a mature CL with a central fluid-filled lacuna (lac).

FIG. 1.10 Transverse transrectal ultrasonographic image of a uterine horn showing moderate endometrial oedema during oestrus.

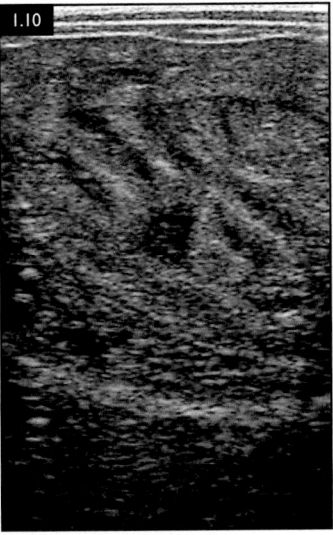

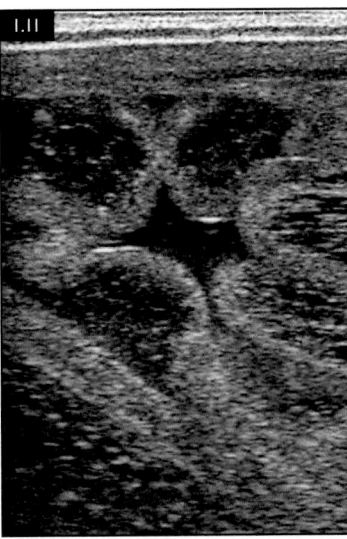

FIG. 1.11 Transverse transrectal ultrasonographic image of a uterine horn showing marked endometrial oedema and intraluminal fluid. Oedema of this degree may be considered excessive.

- predicting ovulation is important for:
  - timing of mating/covering.
  - timing of insemination (frozen semen) crucial if good pregnancy rates are to be achieved.
- no single sign that predicts when a follicle will ovulate, but the following criteria are of use:
  - **Size of follicle.**
    - ◆ follicles grow at 3–5 mm/day during oestrus.
    - ◆ few follicles ovulate when <33 mm in diameter.
    - ◆ size of the follicle at ovulation is NOT determined by the size of the mare.
    - ◆ some mares will routinely ovulate off larger (50–60 mm) or smaller (25–30 mm) follicles.
      - – good record keeping is essential if these mares are to be identified.
  - **Change in consistency of ovulating follicle detected by manual palpation:**
    - ◆ changes from a tense, tight sphere to a soft, fluctuant structure.
    - ◆ 85–90% of pre-ovulatory follicles.
  - **Change in shape of ovulating follicle detected by ultrasound examination:**
    - ◆ round shape to a pear or conical, 'pointing' shape indicates impending ovulation.
    - ◆ 85–90% of pre-ovulatory follicles.
  - **Change in endometrial oedema detected by ultrasound examination:**
    - ◆ maximal at around 1–2 days prior to ovulation.
    - ◆ starts to decline in prominence around 24 hours prior to ovulation.
- type of breeding used will dictate the period of time available for covering or insemination.
  - following timings may be used as a guide:
    - ◆ natural cover – 24–48 hrs. prior to ovulation.
    - ◆ fresh artificial insemination (AI) – 24–48 hrs. prior to ovulation.
    - ◆ chilled AI – 12–24 hrs. prior to ovulation.
    - ◆ frozen AI – 12 hrs. prior to ovulation to 6 hrs. post ovulation.

# Breeding soundness examination of the mare (BSE)

- performed in the mare to determine a mare's suitability as a broodmare:
  - identify causes of infertility.
  - some breed societies standard practice prior to sale.
  - certain parts of the examination are routinely performed during the breeding season as part of the monitoring of the appropriate breeding time in normal mares.

- indications for a BSE include:
  - age: >12 years of age.
  - prior to undertaking expensive AI:
    - frozen semen and embryo transfer programmes (donors and recipients).
  - repeat breeder.
  - repeated early embryonic death.
  - after a problem foaling.
  - mares with a history of endometritis/metritis.
  - Pre-purchase examination of a breeding mare.

## Approach to a breeding examination

### Restraint

- protection of involved personnel is important.
- mare should be suitably restrained, preferably in stocks:
  - nose twitch, other techniques or even sedation may be necessary in some mares.
- tail should be either held out of the way or bandaged.

### General history

- **Age:**
  - young mares may be immature.
  - older maiden mares may have reduced fertility.
  - older pluriparous mares may have 'wear and tear' of their reproductive tract.
- **Type of work/exercise/level of performance.**
- **Previous medical/surgical/lameness.**
  - can the mare carry a foal to term?
  - orthopaedic injuries or evidence of endocrine disease?

## External genitalia

- three barriers maintain the integrity of the reproductive tract:
  - vulva    - vestibule    - cervix.
    - any defect in these increases the risk of uterine contamination and reduces fertility.
- **Vulva**
  - assess for signs of discharge and the position/angle of vulva (Figs. 1.12–1.14).
  - place gloved finger on the pelvic brim and assess the position of the dorsal commissure with respect to the pelvic brim.
    - above the pelvic brim or slopes forward:
      - air can be aspirated (pneumovagina).
      - urine can migrate forward (urovagina) causing uterine inflammation.

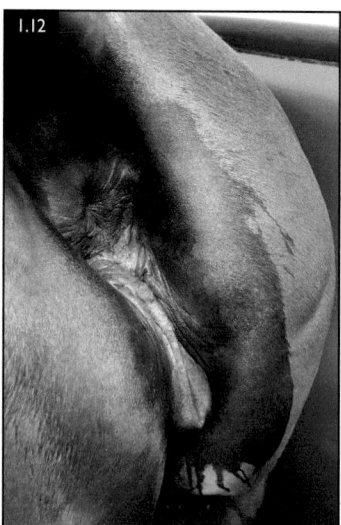 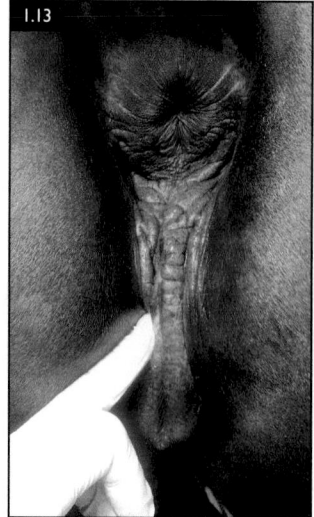 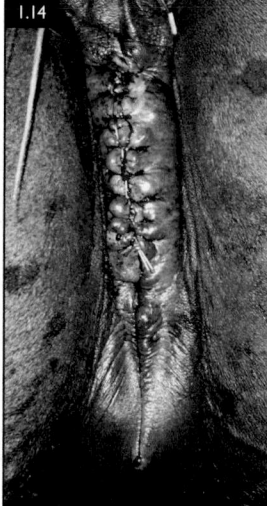

**FIGS. 1.12–1.14** Assessment of the vulval conformation of the mare is an essential part of the breeding soundness examination. (1.12) This Thoroughbred mare has a sloping vulva, with cranial migration of the anus. (1.13) Sloping vulva with finger at pelvic brim for determining the level for the Caslick operation. (1.14) Caslick performed to level of pubic bone.

- o  vulval anatomy can deteriorate with age, following foaling or foaling injury, or loss of body condition.
  - o  **Vulval conformation (VC) good.**
    - ♦  vulval lips are vertical.
    - ♦  lips are closely apposed, making a good seal.
    - ♦  dorsal commissure is either below the bony brim or up to 2.5 cm above.
  - o  **VC fair.**
    - ♦  vulval lips inclined up to 30° off the vertical.
    - ♦  lips are apposed and making a good or reasonable seal.
  - o  **VC poor.**
    - ♦  vulval lips are inclined more than 30° off the vertical.
    - ♦  poor or ineffective seal.
    - ♦  more than half of the vulval commissure is above the bony brim, with a sunken anus.
  - o  vulval lips should be parted to assess:
    - ♦  vestibulovaginal sphincter (no air should enter the cranial vagina).
    - ♦  colour and moisture of the lining of the vestibule.
  - o  integrity of the perineal body is assessed by:
    - ♦  placing one finger in the rectum and the thumb in the vestibule.
    - ♦  at least 3 cm of muscular tissue should be present between the two digits.

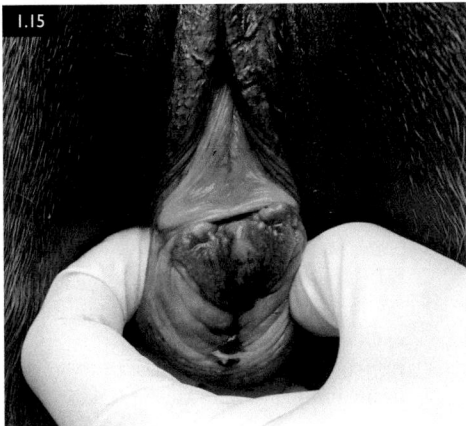

**FIG. 1.15** Clitoral body. Note the fossa and sinuses; both areas should be swabbed during a pre-breeding examination.

- • **Clitoris**
  - o  clitoral fossa and sinuses can harbour venereally transmissible organisms:
    - ♦  *Taylorella equigenitalis* (contagious equine metritis (CEM)).
    - ♦  *Pseudomonas aeruginosa.*
    - ♦  *Klebsiella pneumoniae.*
  - o  special small fine cotton swabs are used to sample the sinuses (Fig. 1.15):
    - ♦  swabs placed immediately into specific bacterial transport media before dispatch.
    - ♦  specialist culture (including microaerophilic) or PCR.
    - ♦  **some countries use a code of practice for CEM and other bacterial venereal diseases, and in some CEM is a notifiable disease.**
- • **Vaginal/cervical examination**
  - o  vaginal speculum or flexible endoscope can be used to inspect the vaginal mucosa and external part of the cervix.
  - o  clean perineal area with plain water and dried with paper towel before insertion of sterile speculum (disposable or re-sterilised). (Fig. 1.16).
  - o  pathology that may be noticed includes:
    - ♦  urine pooling.
    - ♦  uterine or vaginal discharge.
    - ♦  vaginal tears, haematoma, adhesions, masses and varicosities (Fig. 1.17).
    - ♦  persistent hymen.
    - ♦  cervical tears, trauma or adhesions.
    - ♦  rectovestibular/vaginal fistula.
  - o  manual (lubricated) vaginal examination:
    - ♦  assess nature and integrity of the cervix.
    - ♦  **ensure that the mare is not pregnant before dilating the cervix or introducing any instrument through it as will lead to abortion.**

## Rectal and ultrasound examination

- • routine procedure but does carry risk of injury to mare (rectal tear) and personnel involved:
  - o  adequate restraint – stocks and/or sedation.
  - o  always use plenty of obstetric lubrication.

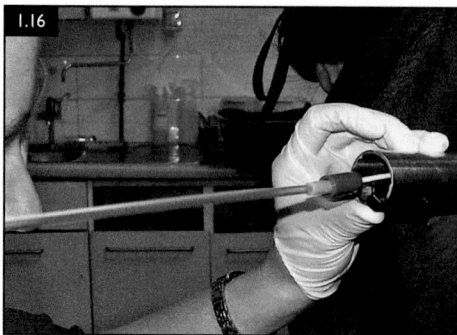

**FIG. 1.16** A sterile vaginal speculum has been inserted to visualise the cervix and vestibule/vagina. A sterile swab is pushed up inside the speculum to take a sample for culture and/or cytology samples from the cervix and uterus.

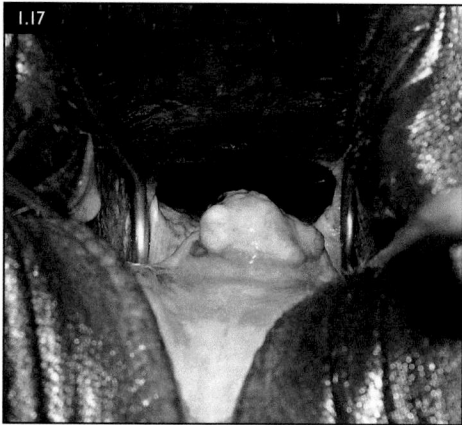

**FIG. 1.17** A view, after opening the vulva and vagina with a speculum, of a ventral vestibular mass, which on removal was found to be benign.

- systematic technique, using 5–7.5 MHz transrectal probe:
  - include examination of left and right ovary, uterine horns, body and cervix.
- record all findings in clinical notes and by images/videos.

## Uterine swab – culture and cytology

- useful for evaluation of breeding mares and detecting uterine infection (culture) and inflammation (cytology):
  - samples taken early in oestrus allow treatment if necessary.
- culture samples usually collected by double guarded swab, via a speculum.
  - plate on blood agar and a selective gram-negative medium (McConkey).

- incubated at 37°C (98.6°F) aerobically for 24–48 hrs. (or 7 days for microaerophilic culture).
  - generally, a pure or heavy growth is likely to be significant.
  - mixed/light growths are likely to be contaminants.
  - potentially pathogenic include:
    - beta-haemolytic *Streptococci*.
    - certain *Staphylococcus* spp.
    - haemolytic *Escherichia coli*.
    - *Pseudomonas* spp.
    - *Klebsiella* spp. (certain capsule types).
    - *Candida* spp.
- cytology samples usually collected by:
  - double guarded swab, via speculum.
  - or small-volume uterine flush:
    - 60 ml of sterile isotonic saline flushed into the uterus, aspirated and centrifuged.
  - sample placed onto a pre-stained slide or slide and stained (Diff-Quik®).
    - examined for presence of neutrophils, endometrial cells and mucus (Fig. 1.18).
- assess cytology and cultures results together (Table 1.2).

## Endometrial biopsy

- useful procedure for:
  - diagnosis of endometrial pathology and prognosis of future fertility in aged mare.
  - mares with a history of pregnancy loss or chronic endometritis.
  - mares assessed as potential embryo transfer recipients (not necessary in every mare).
- equine pathologist experienced in the interpretation of endometrial biopsies is essential to report on varying degrees of inflammation, fibrosis and evaluate glandular changes.
- well established and safe procedure.
  - take during dioestrus but samples can be obtained at any time in the oestrous cycle.
  - inform pathologist of stage of the cycle and other relevant findings/history.
- requires custom-made 70 cm alligator biopsy punch with a basket size of 20 × 4 × 3 mm.

**TABLE 1.2**   Decision making following endometrial culture and cytology.

|  | CULTURE + | CULTURE - |
|---|---|---|
| **CYTOLOGY +** | uterine infection and inflammation. treat uterus in line with bacterial isolate. | uterine inflammation. investigate inflammatory cause. assess vulval conformation. |
| **CYTOLOGY -** | possible contamination of swab. possible false negative if cytology sample was not representative. | no evidence of infection or inflammation. |

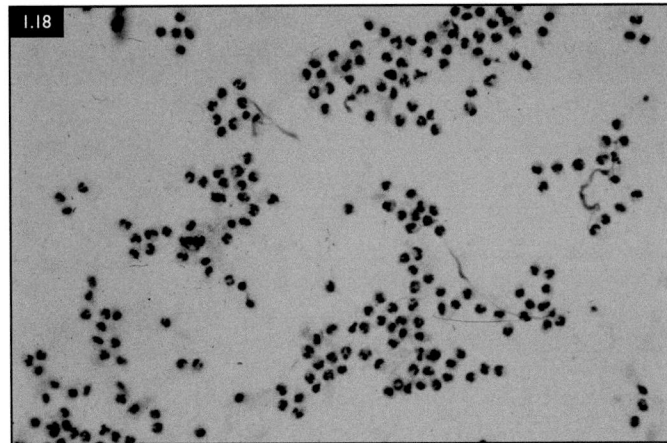

**FIG. 1.18** An endometrial smear from an oestrus mare with large numbers of poly-morphonuclear neutrophils present suggesting the presence of inflammation and possible infection.

- single biopsy from the base of one horn is representative of the entire uterus.
- collected tissue is placed into fixative (formalin or Bouin's solution).
- tissue sections are examined for:
  - inflammation or endometritis
  - periglandular fibrosis.
  - cystic glandular distension and lymphatic stasis (Fig. 1.19).
- classification is based on the severity and distribution of the lesions and is associated with the expected ability of the mare to conceive and carry a foal to term:
  - Category I: slight and/or widely scattered pathology; 80–90%.
  - Category IIA: endometrial changes that reduce breeding efficiency but are moderate or reversible. Inflammation is detected; 50–80%.
  - Category IIB: as for IIA but fibrosis is also present; 30–50%.
  - Category III: irreversible and severe changes of fibrosis, cellular infiltration and lymphatic stasis; 10%.

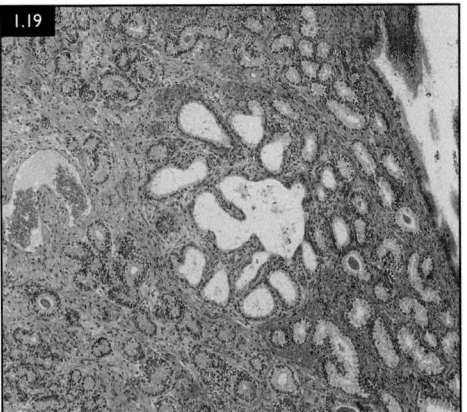

**FIG. 1.19** Histopathology of an endometrial biopsy specimen from a Thoroughbred mare showing an endometrial gland nest, as is often seen in older mares with chronic endometrial pathology.

## Hysteroscopy

- visualisation of endometrium via videoendoscope (sterile) (Fig. 1.20):
  - insufflation with air (or saline) to distend the uterus.

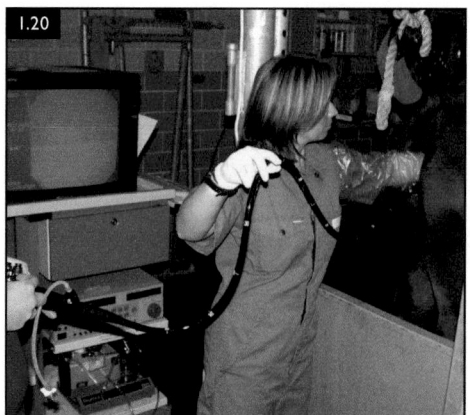

**FIG. 1.20** Examination of a mare by video-endoscopy.

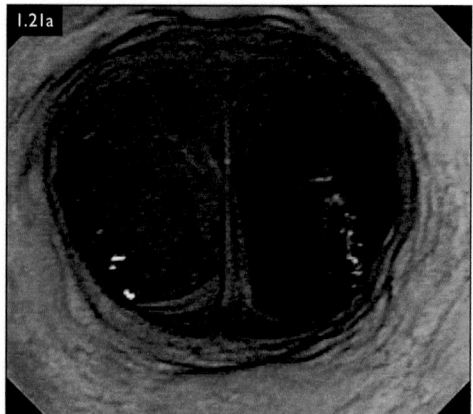

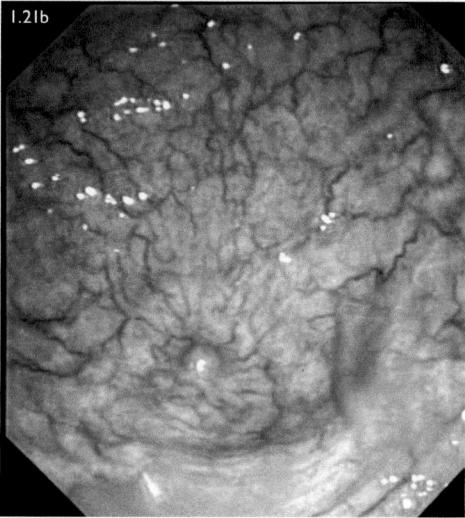

**FIG. 1.21** (a) Hysteroscopic view of the uterine body and separation into two horns of a normal mare. (b) Hysteroscopic image of normal endometrium and uterine papillae.

- ○ perform in dioestrus to help maintain uterine distension.
- can cause mild endometrial trauma/inflammation:
  - ○ perform any other examination including ultrasound prior to endoscopy.
- requires two people to perform; one to pass the scope, one to 'drive'.
- allows assessment of:
  - ◆ vestibule.
  - ◆ external and internal cervical os.
  - ◆ endometrium (Fig. 1.21a).
  - ◆ endometrial cysts.
  - ◆ endometrial damage, adhesions, haemorrhagic foci and polyps.
  - ◆ guiding specific site biopsies.
  - ◆ uterine papillae (Fig. 1.21b).
- transendoscopic insemination:
  - ○ performed using a sterile catheter via the biopsy channel to deposit a low dose of semen on the uterine papillae.
  - ○ helpful in mares susceptible to severe post-covering endometritis and where single straw frozen semen doses are provided.

### Blood endocrine assays

- measurement of blood hormone concentrations is a routine part of broodmare management.
- **Progesterone**
  - ○ detecting presence of functional luteal tissue within the ovary.
  - ○ assessing mares during the transitional period for the first ovulation.
  - ○ guides the use of luteolytic drugs.
  - ○ early pregnancy levels are useful to decide when endogenous progestagens are insufficient (<12.7–15.9 nmol/l [4–5 ng/ml]).
    - ◆ need for supplementation with exogenous progestagens (e.g. altrenogest).
- **Testosterone and inhibin blood levels**
  - ○ may be raised in ovarian granulosa (thecal) cell tumours (GCTs) (See page 70).
- **Oestrone sulphate levels**
  - ○ increase from day 35 of pregnancy.

- o predominately secreted by the foetus and foetal membranes.
- o used as an indicator of foetal viability in the mid–late-term pregnant mare (See page 26).

### Karyotyping

- performed on animals suspected of a chromosomal abnormality.

- o heparinised blood samples are used to harvest lymphocytes:
    - ♦ chromosomal analysis is carried out at specialised laboratory.
- o indicated in mares that have:
    - ♦ never cycled.
    - ♦ infantile/underdeveloped reproductive tracts.
    - ♦ fail to maintain a pregnancy.

## MANIPULATION OF THE FEMALE REPRODUCTIVE CYCLE

### Advancing the onset of Spring Transition

- transition from winter anoestrus into normal cyclicity of the breeding season usually occurs in the early spring.
- can present problems if the mare belongs to one of the breeds (i.e. Thoroughbred) that have an artificial birthday for all foals:
    - o January 1st (northern hemisphere) or August 1st (southern hemisphere).
    - o breeders will strive for early-in-the-year foals.
- early seasonal cyclic abnormality can lead to:
    - o wasted time and resources.
    - o increased workload of the stallion, especially later in the breeding season.
    - o increased veterinary involvement and expenditure.
- breeders often request techniques that can advance the earliest date of mating in barren and maiden mares.

### Artificial lighting

- day length is an important stimulus for the oestrous cycle of the mare.
    - o provision of artificial lighting is the best method of advancing the date of the first ovulation.
    - o several methods have been described to offer the mare artificial lighting conditions.
        - ♦ **24 hours of lighting is detrimental.**
- 16-hour light period and 8 hours dark starting in early December (northern hemisphere) (early June in southern hemisphere).
    - o advances the first ovulation by around 60–80 days.

- o one 150W clear bulb per 16 m² is required.
    - ♦ dark corners of the stable sufficient light to comfortably read a book.
- adding 2–3 hours of light at the end of the natural daylight (as dusk falls).
    - o effective, but mares must be kept outside to receive maximum natural light.
- 'Flash' or 'pulse' system.
    - o delivers 1 hour of light 9.5 hours after the onset of darkness.
    - o not widely used commercially.
- Head collar light (Equilume™).
    - o utilises a specific wavelength blue light (popular in Thoroughbred stud practice).
    - o allows light stimulation and encourages early cycling while being kept at grass and avoiding the expense of stabling (Fig. 1.22)

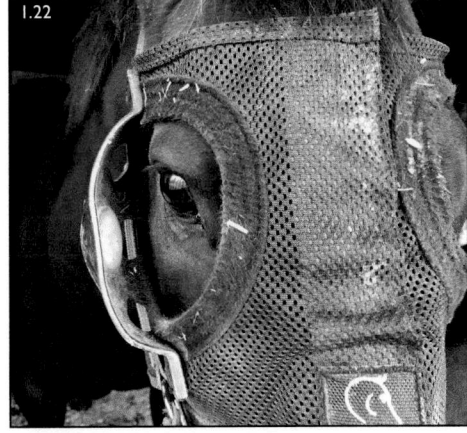

**FIG. 1.22** Mare wearing an 'Equilume' to encourage early oestrous activity through light simulation.

## Mare management

- all barren and maiden mares examined early in the year per rectum for an active CL:
  - blood samples can be taken for plasma progesterone levels to confirm the presence of any active luteal tissue.
- encourage early cyclicity in those mares with some follicular activity and low progesterone levels <3.18 nmol/l (1 ng/ml):
  - treated with exogenous hormones e.g. progesterone withdrawal therapy or GnRH.
- mares with luteal tissue are cycling and can either be:
  - induced into oestrus by exogenous prostaglandin therapy.
  - monitored for the onset of natural oestrus.

## Hormonal methods

- hormonal treatment regimens can be added to the end of the artificial lighting period to further advance the first ovulation date.
- **Progesterone**
  - requires the mare to be in the transitional phase.
  - oral progestagen (0.044 mg/kg altrenogest) daily for 10–15 days.
  - daily intramuscular injections of 150–200 mg of progesterone in oil +/- 10 mg oestradiol 17β for 10 days.
  - intravaginal progesterone-releasing devices for 10 days.
    - vaginitis commonly seen with these devices so do not use in susceptible mares.
  - mare will show signs of oestrus for 5 days (3–6 days) and will ovulate 7–15 days after cessation of progestagen treatment.
    - due to rebound of LH levels that allows final follicle maturation/ ovulation.
  - some clinicians combine this regime with prostaglandin injection on the last day.
    - ensures any luteinised tissue in the ovary will be lysed and ovulation occurs.
    - unnecessary if ovary is monitored by ultrasound and no luteal tissue is noted.

- **Dopamine antagonists**
  - sulpiride (1 mg/kg p/o q24h) and domperidone (1.1 mg/kg p/o q24h) have been used to hasten the onset of the first ovulation in experimental studies.
  - useful in the post-foaling mare that is slow to cycle.
  - also used to stimulate milk production in the post-foaling mare:
    - often used in the 'lactational anoestrous' mare.

- **Synthetic GnRH**
  - GnRH administration results in release of endogenous LH and FSH:
    - most successful method of inducing ovulation early in the transition period.
  - dosing methods include mini-pumps, implants and injections.
    - January to March (northern hemisphere).
    - July to September (southern hemisphere).
  - Buserelin (2.5–3 ml i/m q8–12h for up to 10–14 days).
    - regular scanning (2–3 days) to monitor follicle development.
    - follicle >30–35 mm present, the mare can be teased, and covered.
    - treated with human chorionic gonadotropin (hCG) to encourage ovulation.
    - pregnancy rates following these oestrus periods are around 50%
  - GnRH is available as an implant Ovuplant®
    - implanted into the mare every 48 hours until she ovulates (2–3 implants).
    - more successful if a larger follicle (30 mm) is already present.
    - excessive use can push the mare back into an anovulatory state.
  - human goserelin acetate implants (Zoladex®, 1.8 mg) used subcutaneously:
    - effect limited to 7–10 days.
    - mare requires monitoring every 24–48 hours to identify any ovulatory follicle and then bred.
  - mares can slip back into an anovulatory state after an ovulation with GnRH.

## Synchronisation and manipulation of oestrus during the breeding season

- manipulation of the mare's cycle requires a thorough understanding of physiology:
  - make efficient use of the stud and veterinary staff.
  - synchronise mares in embryo transfer programmes.
  - allow the use of chilled semen on specific days.
- basic approach involves either extending or terminating the luteal phase of the cycle.

## Progestagens

- mimic and extend the luteal phase when administered.
- inhibit behavioural oestrus (stop exhibiting oestrus behaviour after 2–3 days of treatment).
- may lower LH levels and block final maturation of follicles and therefore ovulation.
  - not 100% effective in the mare.
- Note:
  - progestagens do not interfere with the release of $PGF_2\alpha$ from the endometrium:
    - treatment should continue independently of endogenous progestagens for 14–15 d.
  - mares may ovulate while on progestogens:
    - new CL formed will be present when exogenous progestagens are stopped.
    - mare will not return to oestrus until the new CL regresses.
      - $PGF_2\alpha$ given at end of the progestagen treatment period prevents this.
        - ◊ treatment period can be shortened to 8–12 days without any effect on the synchrony of the mares.
    - ultrasound examination prior to and on the last day of exogenous progestagen treatment will allow identification of any CL and, therefore, whether any $PGF_2\alpha$ is necessary.

- some mares develop large follicles towards the end of the progestagen treatment period.
  - these follicles may ovulate rapidly after cessation of the treatment (2–4 days).
  - mares that ovulate at this time may not show oestrus and therefore the opportunity to breed from them may be missed.
  - scanning mares on the last day of treatment to identify those mares with follicles >35 mm in diameter may identify mares that could ovulate early after synchrony.
  - synchrony of mares may be improved by administering hCG to mares that have follicles >30 mm in diameter (see below).

## Prostaglandin F$_2$ alpha (PGF$_2\alpha$)

- causes lysis of the CL, resulting in a decline in serum progesterone.
- CL only regress in response to exogenous $PGF_2\alpha$ more than 5 days after ovulation.
- mare returns to oestrus at 3–4 days after injection:
  - ovulation generally occurring between 5–12 days after injection.
  - administration to mares on day 12–14 may not shorten the time to the next ovulation.
- mares with follicles >35 mm at the time of $PGF_2\alpha$ administration may:
  - ovulate very quickly, or the follicles may slowly regress.
  - using a low dose of $PGF_2\alpha$ in these mares can encourage them to:
    - come slowly into oestrus.
    - allow the uterus and cervix to develop oestrus characteristics.
    - allow mares to be bred before they ovulate.
- some large follicles present at time of $PGF_2\alpha$ may regress.
  - development of a new dominant follicle takes several days.
  - results in an apparent slow response to the $PGF_2\alpha$.
- $PGF_2\alpha$ can also be given to aid synchrony of mares, as described below.

## Combination of progestagen and oestrogen

- oestrogen has a suppressive effect on FSH release.
- combination of progestagens and oestrogen have increased negative feedback on LH than progestagen alone.
- intramuscular injection of 150 mg progesterone in oil and 10 mg oestradiol-17β in oil daily for 10 days, with $PGF_2\alpha$ given on the 10th day, can result in a tight synchrony of oestrus and ovulation.
  - 90% of mares treated ovulate between 10 and 12 days after the last treatment.
- **oestradiol-17β and progesterone in oil may not be currently available commercially in some countries.**
  - other oestrogens are available (oestradiol cypionate and oestradiol benzoate):
    - have different durations of activity.
    - cannot be recommended for use with the above regimen.

## Induction of ovulation during oestrus

### Human chorionic gonadotropin

- luteinising-like effect which helps mature and ovulate a dominant follicle.
- given when the mare is in oestrus with a follicle >35 mm.
- 1500–3000 IU i/m or i/v – 90% of mares ovulate within 48 hours of injection.
- reports of mares that are repeatedly treated with hCG producing antibodies to injectable hCG, causing the mare to fail to respond to the drug – rarely encountered in practice.

### GnRH implant/injections

- Deslorelin preparations include:
  - Ovuplant® commercially available biodegradable, silastic implant.
    - injected into mare's neck or vulval tissue when oestrus mare has follicle >30 mm.
    - response similar to hCG, but with no risk of stimulating an antibody response.
    - reported increase in the intraovulatory period can be avoided by inserting the implant into the vulval mucosa and removing it following ovulation.
  - injectable deslorelin is available in some countries and is used in a similar way.

## Managing the cyclicity of the post-partum mare

- Foal heat is the first heat period following parturition at 5–12 days.
- ovulation usually occurs 8–11 days after foaling (shortens later in the breeding season).
- chance for the post-partum uterus to involute, recover and clean itself.
  - good opportunity to palpate and scan the mare and collect endometrial swabs.
- some breeders are keen to mate the mare at foal heat (earlier foal the following year).
- pregnancy rates are higher if a mare ovulates >10 days after foaling compared with a mare that ovulates <10 days after foaling.
  - histologically the endometrium takes approximately 14 days to repair following foaling and return to a normal 'non-pregnant' state.
  - with the embryo entering the uterus 4–5 days after ovulation the day 10 cut off allows the uterus to be fully recovered.
  - mares in foal from being bred at foal heat have a slightly increased pregnancy failure rate when compared with mares bred at the second oestrus or those that are short cycled.
- mares for possible foal heat cover should be examined at 7 days post foaling by vaginal speculum, rectal palpation and ultrasound:
  - select those capable of being bred on the foal heat.
  - **better to short cycle the mare than to compromise on a foal heat cover.**
- criteria used to decide whether to cover a mare at foal heat include:
  - normal foaling.
  - no retained placenta or history of metritis or endometritis.
  - no cervical, vaginal or vulval damage.

○ normal uterine involution.
○ ovulation > 10 days post foaling.
- intensive post-breeding treatment may be needed to improve pregnancy rates at foal heat.
  ○ intrauterine antibiotics, intravenous oxytocin +/- uterine lavage.

## PGF$_2\alpha$ administration

- can be used to shorten the first luteal phase in mares that were not ideal for mating on or around day 10.
- can be given to mares 5 or 6 days after ovulation when the exact dates are known, or on or around day 20 post foaling to bring them back into oestrus 3–4 days post injection.

## Delaying the first post-partum ovulation

- progesterone or altrenogest alone (or in combination with oestradiol) can be used to delay the first ovulation.
- can be given from day 1 post-partum but can interfere with uterine health recovery:
  ○ preferable to start from around day 4 to allow some uterine involution to occur.
  ○ delaying the first ovulation to beyond day 10 improves pregnancy rates for mares bred at foal heat but this technique is not always successful.

---

## BREEDING MANAGEMENT

## Minimal-contamination breeding techniques

### Overview

- during natural mating ejaculation occurs into the body of the uterus.
- with AI (fresh, chilled or frozen) a catheter is passed through the cervix and semen deposited in the uterine body:
  ○ uterine tubule for frozen semen by the deep intrauterine technique.
- both techniques can lead to contamination of the uterus and inflammation of the cervix, vestibule, vagina and vulva:
  ○ natural covering more than artificial techniques.
- physiological uterine reaction to the semen regardless of the breeding method:
  ○ results in intrauterine inflammatory fluid expelled within 12–24 hours of breeding.
  ○ persistence of the fluid is pathological and requires treatment:
    ♦ post-covering examinations are therefore advised.
- techniques can be applied at mating aimed at reducing the pathogen challenge to, and any inflammatory reaction from, the intrauterine environment in the pericoital period.

○ help to improve fertility rates in:
  ♦ older multiparous mares.
  ♦ those with a history of endometritis.
  ♦ mares mated at the foaling heat.
  ♦ mares mated to stallions infected with *Pseudomonas aeruginosa* or *Klebsiella pneumoniae* (not in the UK under the HBLB Code of Practice).
- **considered sound 'good practice' in any mating situation to minimise the possibility of infection and maximise fertility.**

### Technique

- one covering per oestrus period:
  ○ timing of breeding dependent on stallion choice/semen type.
  ○ use of ovulation agents, LH or deslorelin, can help coordinate breeding/ovulation.
- general hygiene at covering:
  ○ bandage the tail.
  ○ wash the vulva and perineum of the mare (possibly wash the penis of the stallion).
    ♦ clean water, preferably from a spray bottle or from a disposable plastic liner in a bucket (plastic bin liner) to avoid cross-contamination.
    ♦ dry with a sterile paper towel.

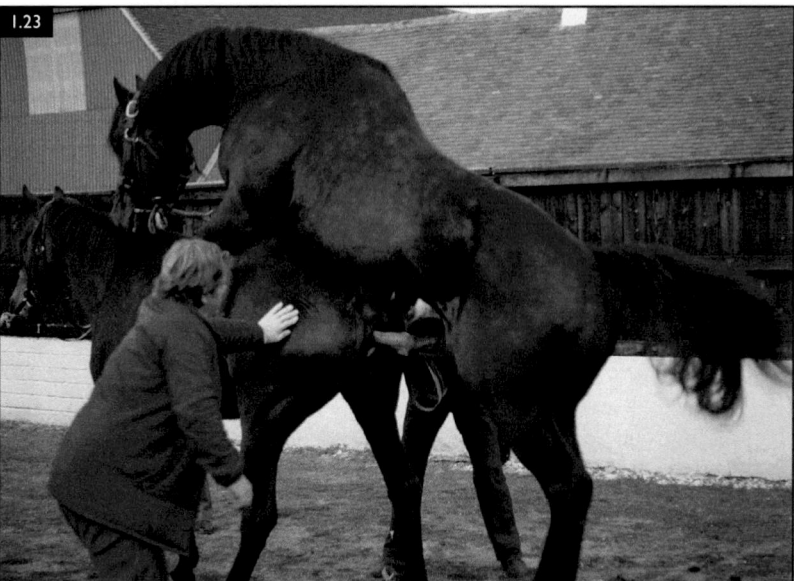

**FIG. 1.23** A natural covering in a Thoroughbred mare.

- ♦ do not use chemicals or strong soaps/detergents.
  - – can be spermicidal.
  - – alter normal bacterial population of reproductive tract leading to pathological bacterial overgrowth.
- may be necessary to use warmed semen extender with antibiotic into the uterus at covering.
- mating should be fully supervised.
- assess presence of any intrauterine fluid 4–24 hours after mating by ultrasound examination:
  - ○ significant quantity of fluid present after 4 hours, the uterus should be lavaged:
    - ♦ 1 litre lactated Ringer's solution or 0.9% saline warmed to 37°C [98.6°F].
    - ♦ via an equine uterine lavage catheter or soft tube:
      - – lavage/siphoning repeated a maximum of 3–4 times or until the flushing solution emerges clear of debris.
  - ○ pre-breeding endometrial swab shows bacterial contamination:
    - ♦ antibiotic solution may be infused after the last flushing depending on the culture and sensitivity results.

- ♦ routine use of intrauterine antibiotics post breeding is controversial and not best practice.
- Oxytocin administered im. or iv. 4–8 hours after mating, has proven beneficial to conception rates in some mares with a history of chronic endometritis.
  - ○ after i/v oxytocin wait 30 minutes until myometrial activity subsides before instilling an antibiotic solution.
  - ○ re-examine mare by ultrasound 24 hours later and repeat oxytocin injection every 4–8 hours if required.
- uncommonly, the mare may require further uterine lavages, particularly in those animals with pre-existing issues such as cervical fibrosis or poor lymphatic drainage.

## Natural mating

- mare is physically mounted and bred by the stallion.
- carefully supervised breeding where both stallion and mare are handled in a managed environment, the mare is bred once, at a precise time of the oestrus period:
  - ○ common practice in Thoroughbred breeding (Fig. 1.23).

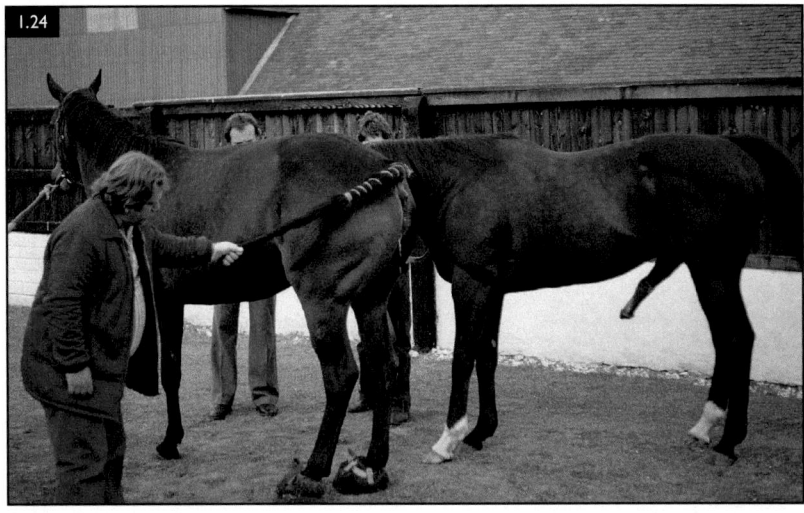

**FIG. 1.24** Teasing of a Thoroughbred mare by a stallion prior to natural covering. Note the restrained mare with a nose twitch and padded boots on the hind feet. The stallion is presented to the mare from the side to minimise the chances of being kicked.

- o requires accurate ultrasound monitoring of the reproductive tract with/without a teaser stallion (see below).
- o ovulation is detected by ultrasound 24-28 hrs. after breeding.
- o pregnancy scan 14 days after ovulation.
- paddock breeding where mares are turned out with the stallion and left to breed multiple times during the oestrus period:
  - o ovulation is assumed to have occurred and the mare 'in foal' if she doesn't return to oestrus 16-18 days later.
  - o pregnancy scan should be performed at 16-18 days after breeding to detect twin pregnancy and again at 21-28 days.

## Use of a teaser stallion

- teaser stallion must have the same pre-breeding health tests as mares/stallion:
  - o CEM (urethra, urethral fossa, sheath/prepuce, pre-ejaculatory fluid), equine viral arteritis (EVA) and EIA.
- requires an experienced team to handle the teaser and mare and make accurate observations.
- teasing can be performed over a 'teasing board' where a solid gate/wall separates the two horses to minimise the risk of kicking/injury but allows head-head and stallion head-mare tail contact.
- 'Open teasing' where no barrier is used can be a more effective method but requires increased risk management (Fig. 1.24).
- teasing is performed every 1–2 days and records made of the mare's response.
- allows mares to be presented to the veterinary surgeon for ultrasound examination when 'in oestrus' thus reducing the number of examinations and costs.
- some breeding operations do not use veterinary monitoring of the mares:
  - o rely only on how the mare behaves towards the stallion.
  - o can be labour intensive and require multiple matings per oestrus period.
- 'in oestrus' mare can be repeatedly mounted (but NOT mated) by the teaser.
  - o confirm mare will stand to be covered.
  - o familiarises the maiden mare with the process and reduces the stress of the procedure.
- teasing after covering can help evacuate any uterine fluid (should not be performed less than 6 hours post covering).

# Artificial insemination of the mare

- regularly used in many types of horse and pony to achieve a pregnancy:
  - some breeds, however, will not register progeny bred by this technique.
- three types of insemination are used in the horse:
  - fresh +/- extender.
  - chilled extended.
  - frozen extended.
- technique and technology of semen collection, extension, preservation, monitoring and transport are covered in (See page 116).
- **advantages of AI are:**
  - increased conception rate in some stallions and mares with fresh or chilled semen.
  - mares can be bred by stallions that would otherwise be geographically inaccessible.
  - mares and foals are not subject to transport stress.
  - mares and stallions can remain in training.
  - mare owners save on transport costs.
  - reduced contamination of mares and decreased risk of disease transmission:
    - control/prevention of venereal diseases.
  - reduced risk of injury to mare and stallion as there is no physical contact.
  - increased flexibility in timing of mating:
    - maximises the use of a stallion during and beyond his reproductive years.
  - single semen collection split between multiple mares:
    - reduces workload of the stallion.
- additional benefits of frozen semen include:
  - semen can be kept indefinitely.
  - mare can be bred at any time, including when the stallion is unavailable.
  - worldwide availability of semen.
  - single shipping of semen per season.
- **disadvantages of AI are:**
  - veterinary input and expense are greater, especially when using frozen semen:

- specialised equipment and laboratory facilities required.
  - communication between veterinarian, mare owner and stallion owner are essential.
  - repeat inseminations lead to increased semen collection and shipping costs.
  - conception rates using frozen semen may be less than with fresh/chilled semen or natural covering:
    - good management, experience, modern techniques and ovulatory drugs, conception rates are comparable.
    - huge variation between and within stallions in:
      - ability of their semen to withstand freezing and thawing techniques.
      - longevity of the chilled semen.
  - international custom controls:
    - quarantine requirements for frozen semen processing and collection.
    - specific import/export requirements for movement of chilled/frozen semen.
  - reliable courier service is essential for delivery of chilled semen.
- cost of some AI programs may mean that a full BSE is indicated to avoid insemination in a mare that does not have a uterus capable of facilitating conception and pregnancy:
  - recommended that all mares and stallions have appropriate pre-breeding tests performed (EVA, CEM, EIA).
  - cervical/endometrial swab and smear taken in oestrus to identify any endometritis or uterine inflammation.

## Fresh or chilled semen

- ultrasound scan the mare until:
  - in oestrus with >35 mm follicle, uterine oedema and a relaxed cervix.
- liase with the stallion owner/agent well in advance of needing semen to allow any stallion arrangements to be made:
  - order the semen and plan to have the mare inseminated within 24 hours (48 hours maximum) of collection.
- aim to inseminate the mare prior to ovulation and for ovulation to occur within 24 hours of insemination.
  - use of hCG or deslorelin is routine.

- check that appropriate paperwork accompanies the semen.
- inseminate into the uterine body using a syringe without latex.
- check motility of semen on a warmed microscope stage.
- check the next day for ovulation and treat accordingly (e.g. uterine lavage, oxytocin).
- check mare for pregnancy at 14-15 days post ovulation.

## Frozen semen

- optimum time for insemination is between 12 hours prior to ovulation and up to 6 hours after ovulation.
- ultrasound scan the mare:
  - in oestrus, follicle of >35 mm, uterine oedema and a relaxed cervix.
- administer an ovulating agent (hCG or deslorelin).
- scan the mare every 8 hours until ovulation is detected:
  - inseminate the mare with a deep uterine insemination (DUI) technique.
- check sperm motility on a slide on a warmed microscope stage.
- scan the mare 4–8 hours later and treat for uterine fluid with uterine lavage and oxytocin.
- re-examine the mare every 12–24 hours until no uterine fluid remains.
- check the mare for pregnancy at 14–15 days post ovulation.
- various protocols have been developed in order to minimise the number of veterinary examinations:

See text boxes for protocols to reduce examinations and minimise over-scanning.

## Frozen semen insemination procedure (Figs. 1.25–1.27)

- specific storage and handling guidelines to ensure the integrity of straws is maintained and the fertility of the semen is not compromised.
- semen is usually supplied in 0.5–5 ml straws:
  - each straw or packet is individually labelled.
  - thawing should be as per the supplied instructions in a clean

---

- single doses of frozen semen or as a general approach these timings may be used to avoid night-time examinations. Scan mare – 35–40 mm follicle
- **Day 1** Deslorelin given **8 pm** – 0 hrs
- **Day 2** Scan mare (and am) 4–6 pm – 20–22 hrs
- **Day 3** Scan mare and inseminate if ovulated or if follicle close **8 am** – 36 hrs
- **Day 3** Scan mare 4 pm and inseminate if ovulated or treat if inseminated earlier **4 pm** – 44 hrs
- **If mare not ovulated, she should be scanned every 6–8 hrs until ovulation then inseminated.**

---

- Multiple doses of frozen semen available a 'fixed time' protocol can be followed.
- Scan mare – 35–40 mm follicle
- **Day 1** Deslorelin given **4 pm** – 0 hrs
- **Day 2** Scan mare and inseminate *Dose 1* **4 pm** – 24 hrs
- **Day 3** Scan mare even if ovulated and inseminate *Dose 2* **10 am** – 42 hrs
- **If mare not ovulated, scanned every 6 hrs until ovulation and a third insemination given.**

---

temperature-controlled water bath or polystyrene box containing water.
  - take great care to avoid semen contact with disinfectants, detergents or tap water and avoid temperature fluctuations – compromises the sperm.
  - once thawed the straw temperature should be maintained and inseminated immediately in its entirety.
- mare to be inseminated should be prepared in advance as described below.
- frozen semen should ideally be inseminated by the deep intrauterine insemination (DUI) technique which gives improved pregnancy rates.

## Preferred insemination process

- cut the 'crimp end' off the straws, take care to preserve all the semen.

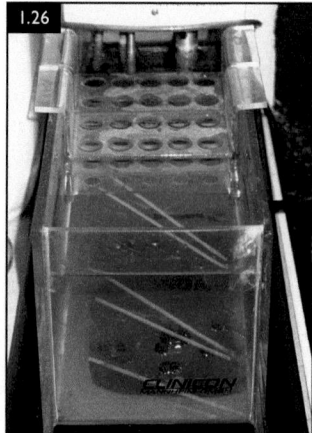
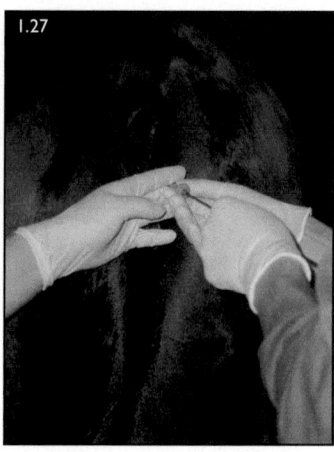

**FIGS. 1.25–1.27** (1.25) The straws containing the frozen semen should be stored in liquid nitrogen and carefully labelled to avoid incorrect insemination. (1.26) The straws are thawed in a temperature-controlled water bath before insemination. (1.27) Insemination of a mare with frozen semen using an insemination pipette.

- insert the catheter through the cervix into the body of the uterus.
  - hand should be placed rectally to guide the catheter up the uterine horn on the side of the ovulation.
- insert the cut end of the straw into the DUI catheter:
  - use the insemination gun/rod to push the cotton plug through the straw and deposit the semen in the uterine tip.
- remove the insemination rod, pull off the used straw and repeat the process until all straws are used.
- small volume of semen is often left by the cotton plug and can be assessed on a warmed microscope slide for sperm motility.

## Alternative insemination process

- cut the 'crimp end' off the straws, take care to preserve all the semen.
- hold the cut end over a pre-warmed (37°C), sterile conical container.
- cut the plug end off the straws to allow the semen to collect in the conical container.
- use an all-plastic syringe preloaded with 5 ml of air to draw the thawed semen into a sterile pre-warmed insemination pipette.
  - inseminate into the body or, ideally, DUI.
  - preloaded air ensures all the semen is inseminated.

- any remaining semen can be assessed on a warmed microscope slide for sperm motility.
- once thawed, the semen should never be refrozen.
- generally recommended that there is a minimum of 30% progressively motile sperm after thawing:
  - dosages of $150–350 \times 10^6$ of these motile sperm reported to result in acceptable conception rates.
- insemination of thawed frozen semen has been associated with a severe inflammatory reaction in some mares:
  - requires post-insemination assessment at 6–8 hours with uterine lavage, intrauterine antibiotics and systemic oxytocin treatment as necessary.

## Preparation of the mare for any AI

- empty the rectum.
- tail wrapped in plastic sleeve and held to one side.
- wash perineal region thoroughly with clean, warm water and dry.
- use a clean plastic rectal sleeve.
- use a non-spermicidal lubricant.
- when using fresh or chilled semen, the material is drawn into a non-toxic syringe (latex free) and a sterile insemination pipette.

# Embryo transfer (ET)

## Overview

- allows mares to breed without carrying the foal herself:
  - mare continues competing while producing foals.
  - multiple foals per year from one mare.
  - sub-fertile mares can continue to breed.
  - mares that have sustained an injury that prevent her carrying a foal can still breed.

- ET donor mare is mated (natural or AI).
- uterine flush performed to harvest the embryo.
- either stored using cryopreservation or implanted into a recipient mare:
  - recipient mare is synchronised to the donor mare's cycle.
  - cryopreservation allows the pregnancy to be delayed until the following years and to ensure a foal born at the desired time of year.
    - ◆ success rates may be lower using frozen embryos.

## THE PREGNANT MARE

## Maternal recognition of pregnancy

- first maternal recognition of pregnancy occurs at about 48 hours post fertilisation:
  - production of a protein called 'early pregnancy factor' (EPF).
  - detected in horses, humans, sheep, and mice.
- embryo enters the uterus at around day 5.5–6 days post ovulation:
  - either a late-stage morula or an early stage blastocyst.
- maternal recognition of pregnancy prevents the production of $PGF_2\alpha$:
  - movement of embryo around the uterus allows contact to every part of the endometrium.
  - secretion by the embryo of anti-luteolysins/luteotropic substances:
    - ◆ taken up by the endometrium.
    - ◆ exact nature is not fully known.

## Endocrinology and maintenance of pregnancy (Fig. 1.28)

## Progesterone

- essential for the maintenance of pregnancy:
  - primary CL forms from the ovulation that led to the pregnancy.
    - ◆ persists because of the inhibition of $PGF_2\alpha$ secretion.

- ◆ after days 35–40 of pregnancy:
  - – eCG from the endometrial cells stimulates additional follicles on both ovaries.
  - leads to secondary CL:
    - ◆ continue to maintain progesterone levels until day 120 when levels decrease.
  - foetoplacental unit begins to secrete progesterone from day 60:
    - ◆ by day 100 producing enough progestins to maintain pregnancy.
- progesterone assays can help in pregnancy diagnosis and assessment of quality of primary CL.
  - **indication of luteal tissue NOT of pregnancy.**
  - elevated progesterone from a mare 18–22 days after the last ovulation suggests an active CL and possible pregnancy.
  - false positives occur due to:
    - ◆ dioestrus ovulation.
    - ◆ embryonic loss after maternal recognition of pregnancy.
    - ◆ failure of luteolysis   ◆ pyometra.

## Equine chorionic gonadatrophin (eCG)

- produced by endometrial cups at the base of the pregnant uterine horn:
  - develop from embryonic cells that invade the endometrium around day 35.
- LH-like effects.

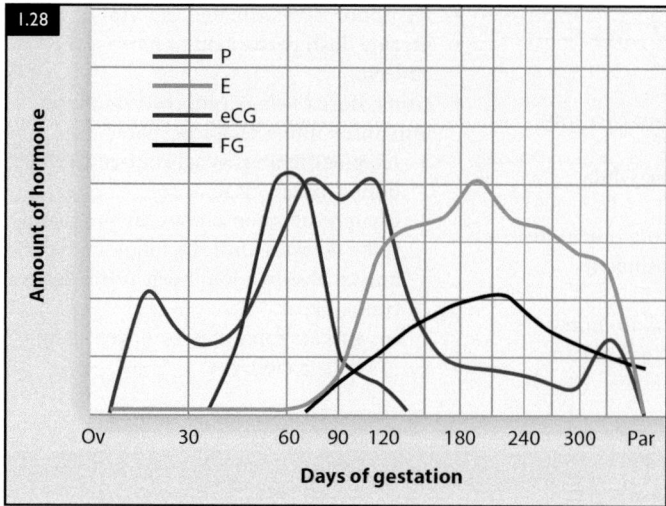

**FIG. 1.28** Summary of hormonal events in the pregnant mare. P = progesterone; E = oestrogen; eCG = equine chorionic gonadotropin; FG = foetal gonad weight. (Adapted from McKinnon AO, Voss JL (1993) (eds) *Equine Reproduction*. Lea and Febiger, Philadelphia, pp. 27–175.)

- ○ stimulates the primary CL.
- ○ luteinisation of secondary CL, maintaining progesterone levels.
- ○ involved in maternal immunotolerance of foreign antigens produced by the foetus.
- basis for a pregnancy test in mares from days 40–42:
  - ○ eCG requires an embryonic source of the cells so only elevated in mares that are/have been pregnant.
  - ○ false positives occur when pregnancy is lost post-endometrial cup formation.
- endometrial cup lifespan is up to 100–120 days of pregnancy:
  - ○ occasionally persist beyond this causing abnormal cycling in next breeding season.

## Oestrogen

- maternal serum oestrogen levels rise from about day 35 of pregnancy:
  - ○ eCG stimulates luteal steroidogenesis – increased oestrogen synthesis and secretion.
- after day 45, additional oestrogens are produced by the foetoplacental unit and released into the maternal circulation:
  - ○ levels peak at around days 210–250 and then slowly decline.
- role in development of vascular supply and endometrial hypertrophy during pregnancy.
- oestrone sulphate can be used as an indicator of pregnancy and foetal viability:

- ○ only produced in the presence of a viable foetus.
- ○ assays can be used from days 100–120 onwards.

## Pregnancy diagnosis in the mare

- essential aspect of veterinary management on the stud farm:
  - ○ various techniques available depending on the stage of pregnancy.
- essential to identify a **single** viable pregnancy and to monitor its development:
  - ○ early embryonic death is common in the mare.
  - ○ leads to a wasted breeding season.
- twin pregnancy (or triplets/quads etc.) must be detected:
  - ○ managed appropriately to prevent serious complications later in pregnancy.

## Absence of oestrus behaviour

- absence of a mare returning to oestrus is an indirect test for pregnancy:
  - ○ **failure to exhibit oestrus behaviour does not mean the mare is pregnant.**
  - ○ other reasons include:
    - ♦ early embryonic death
    - ♦ retained CL
    - ♦ silent or poorly shown oestrus.
    - ♦ variability in the oestrous cycle timing

- ♦ lactational anoestrus.
- ♦ teasing at the incorrect time in relation to the true date of the last ovulation.
- ♦ some pregnant mares show oestrus behaviour.
- ♦ twin pregnancies will be missed.

## Rectal examination

- most frequently used method for pregnancy diagnosis:
  - ○ usually combined with transrectal ultrasound examination.
- carries risks to the mare and veterinarian.
- requires skill and experience for success.
- main findings are noted in Table 1.3 and relate to changes in uterine tone and enlargement.

## Vaginal examination

- direct vaginal examination and viewing of the cervix is **not considered a safe way of confirming pregnancy.**
- risk of inflammation/bacterial introduction which can lead to prostaglandin release, luteolysis and pregnancy failure.

## Ultrasonographic examination

- most important technique for the diagnosis and assessment of pregnancy in the mare:
  - ○ used at same time as rectal examination and therefore shares that technique's inherent risks.
- mare is examined 14 days after confirmed ovulation or 16 days after cover.
- requires a careful, complete and systematic approach:
  - ○ linear 7.5 MHz transducer with the screen out of direct sunlight.
  - ○ essential mare is well restrained.
  - ○ main characteristics of the conceptus are described below:
- **keep accurate clinical records of uterine cysts and the number of follicles at the time of cover or CLs after cover to be alert to the risk of multiple pregnancies.**
- **repeat ultrasound examinations are recommended to monitor normal development of the pregnancy and to confirm the presence of a single pregnancy only.**

| TABLE 1.3 Main findings on rectal examination of a pregnant mare | | | |
|---|---|---|---|
| **GESTATION** | **UTERINE TONE** | **UTERINE SWELLING** | **CERVICAL FINDINGS** |
| 16–19 days | Good, firm/turgid | Mild swelling | Closed tight |
| 20–24 days | Good, firm/turgid | Ventral bulge at uterine bifurcation | Narrow and elongated |
| 30 days | Good, firm/turgid | Increasing pregnant horn ventral enlargement | Closed tight |
| 31–50 days | Decreases in pregnant horn. Still good in non-pregnant horn | Continuing increase in size of bulge. Towards end of period starts to include mid-uterine horn and uterine body | Closed |
| 51–70 days | Decreases in pregnant horn. Still good in non-pregnant horn | Fluid-filled swelling pregnant horn, body, and some distension non-pregnant horn | Closed |
| 71 + days | | Uterus increases in size and moves ventrally. Ovaries start to move closer together. Ballottement of the foetus is possible later in pregnancy (>120 days). Palpation and sizing of foetus may allow ageing by comparison with breed standards | Closed |

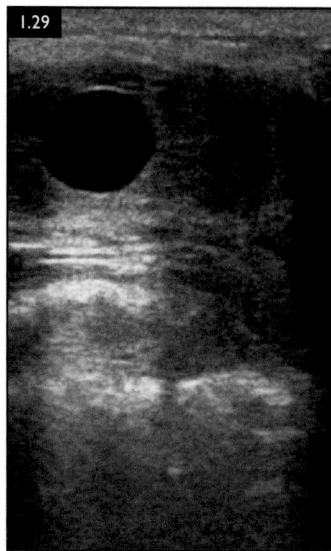

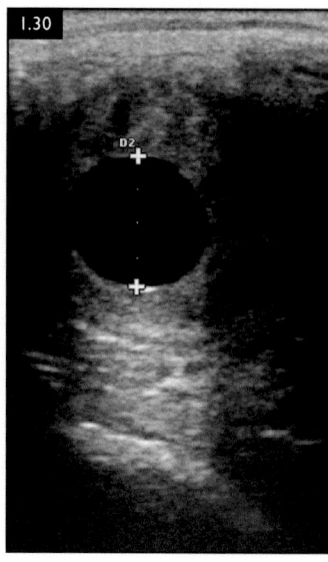

**FIG. 1.29** Transverse transrectal ultrasonographic image of a single 13-day-old conceptus.

**FIG. 1.30** Transverse transrectal ultrasonographic image of a single 15-day-old conceptus.

### Days 10–16 (Figs. 1.29, 1.30)

- entire uterus must be examined as the conceptus is very mobile:
  - found from the tip of the uterine horn to the internal cervical os.
- day 14 (from ovulation) conceptus 14 mm spherical vesicle:
  - bright echogenic poles (6 and 12 o'clock position).
  - not associated with the embryonic disc.
- vesicle grows from about 2–3 mm (day 10 from ovulation) to 14 mm (day 14) to about 15–16 mm (day 16):
  - ideal time to detect and manage twins.

### Days 17–22 (Figs. 1.31, 1.32)

- vesicle has a growth plateau between days 17 and 26:
  - often an irregular/triangular or slightly elongated shape due to increased uterine tone and wall thickness.
  - embryo and its heartbeat first detected on day 21 usually at the 5–7 o'clock position.

### Days 23–55

- day 24 the allantois can be seen:
  - expands as the yolk sac contracts to lift the embryo into the vesicle (Fig. 1.33).
- after day 40, the yolk sac dramatically reduces, and the umbilical cord is formed:
  - allows the embryo to drop down to the ventral portion of the vesicle by day 50.

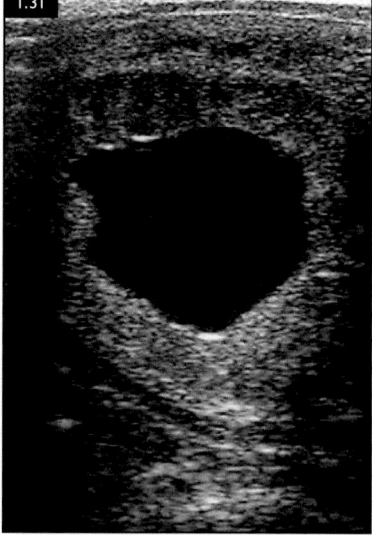

**FIG. 1.31** A 19-day (from ovulation) pregnancy ultrasound scan.

  - **twin vesicle walls, when in contact with one another, generally appear vertical.**
    - usually, horizontal membranes of allantois and yolk sac of a 30-day-old vesicle.

### Days 60–80 (Figs. 1.34, 1.35)

- embryo moves forward and over the pelvic brim:

- makes complete assessment of embryo difficult, especially in large mares.
  - presence of twins at this stage can be very difficult to rule out.
- sex of a foal can be determined from day 55–60 (optimal window day 60–65).
  - genital tubercle develops between the hindlimbs.
  - visible as a highly echogenic, bilobed structure resembling an equal sign (=).
  - migrates caudally in the female to a position under the tail.
  - moves cranially in the male onto the abdominal wall behind the umbilicus.
  - requires a scanner of 5–7.5 MHz frequency – good resolution at 12–15 cm depth.
  - considerable experience and practice to become accurate in this technique.
- abnormalities of the foetus and pregnancy can also be identified:
  - lack of foetal heart beat.
  - cloudy echogenicity or reduction of foetal fluids.
  - rarely, morphological abnormalities.
- second window for foetal sexing is between 100–120 days:
  - foetus 'floats' back up into the pelvis where it can be imaged (much larger).

### Day 180 onwards

- transabdominal ultrasonographic examination is possible at this stage and allows:
  - assessment of the foetus
  - gender determination.
  - placental membrane and fluid assessment.

## Hormonal tests for pregnancy

- indirect testing of pregnancy:
  - measuring a variety of hormones in the mare's blood or milk.

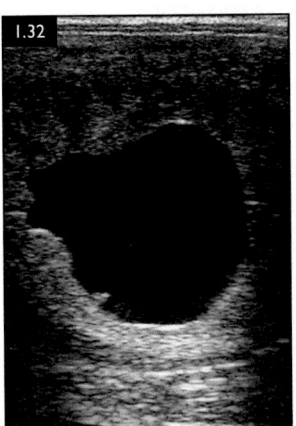
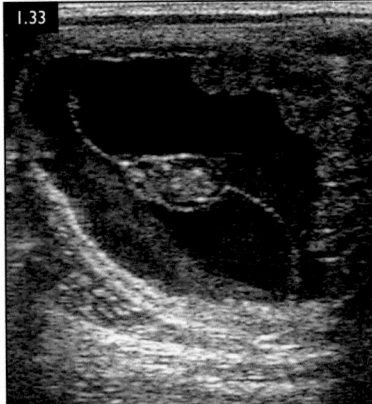
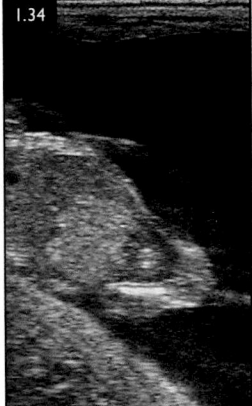

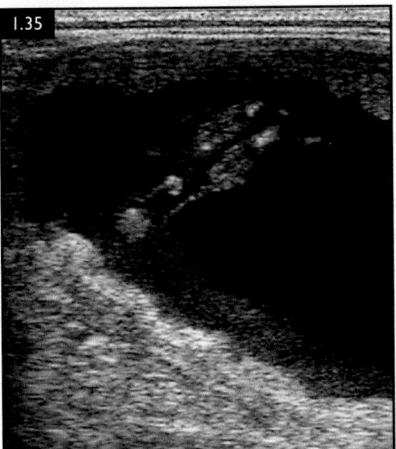

**FIGS. 1.32–1.35** (1.32) 21-day pregnancy with the embryo seen at the 7 o'clock position. (1.33) 30-day pregnancy with the embryo seen in the centre separating the allantoic and yolk cavities (both have equal volume at 30 days). (1.34) 65-day pregnancy scan showing a colt foal. The hyperechoic genital tubercle can be seen in front of the hindlimbs. (1.35) 65-day pregnancy scan showing a filly foal. The hyperechoic area lying between the tail (lower left) and two hindlimbs is the genital tubercle of the filly.

- not reliable and does not rule out twin pregnancies.
- useful in mares that cannot be examined by rectal ultrasound.

## Progesterone

- blood assay
- elevated levels (>2 ng/ml) when active CL present, often 4–10 ng/ml.
- used at 17–24 days post ovulation to show mare not returning to oestrus.
  - **does not mean the mare is pregnant as false positives.**

## Equine chorionic gonadotropin

- produced by endometrial cups from days 35–38 post ovulation.
- levels reduce after day 100 due to maternal rejection.
- false negatives if samples are taken outside these time limits or with mule foetuses.
- false positives with embryonic death after day 35.

## Oestrogen

- blood assay.
- useful indicator of foetal viability:
  - foetal gonad (and the placenta) contributes to production.
- used 100–120 days of pregnancy:
  - levels fall during the final few weeks of pregnancy.

## Techniques for assessing foetoplacental health

- pregnant mare should be closely monitored during the early stages of pregnancy:
  - routine scans at 14, 21/28, 35/42 days after ovulation.
  - foetal sexing is ideally performed by transrectal ultrasound at 60–65 days.
- mid–late-pregnant mare is not monitored to the same extent or at all unless:
  - problem is suspected, or she has a history of pregnancy failure.
  - mare is classed as 'high risk' and warrants regular monitoring:
    - premature mammary gland development.
    - vulval discharge.
    - mare systemic illness.

- post-surgical management.
- history of previous pregnancy failure – abortion/resorption.
- history of premature, dysmature or septic foals.

## Blood analysis

- **Oestrogen.**
  - increased levels at 190–280 days (>1000 ng/ml considered normal) returning to low level at parturition.
  - levels <500 ng/ml prior to 300 days suggest foetal compromise.
  - **changes do not always occur.**
- **Progesterone.**
  - levels increase during the last month of pregnancy.
  - increase prior to 310 days is associated with placental pathology.
  - useful test but requires care in interpretation.

## Changes in mammary secretions

- during the last week of gestation characteristic changes occur in the milk electrolyte levels in readiness for parturition (Fig. 1.36).
  - potassium levels rise.
  - sodium levels fall.
  - potassium and sodium levels cross over approximately 48 hours prior to parturition.
  - calcium rises to over 10 mmol/l.
  - premature mammary development or milk changes occurring prior to 310 days:
    - placental pathology suspected (placentitis, twins) and should be investigated.

## Placental ultrasound

- placenta can be assessed by transrectal and transabdominal ultrasound:
  - requires an experienced examiner and availability of equipment.
- **Transrectal technique:**
  - rectal palpation assessing size of uterus.
  - monitor foetal movement over time:
    - absence of movement is not unusual.

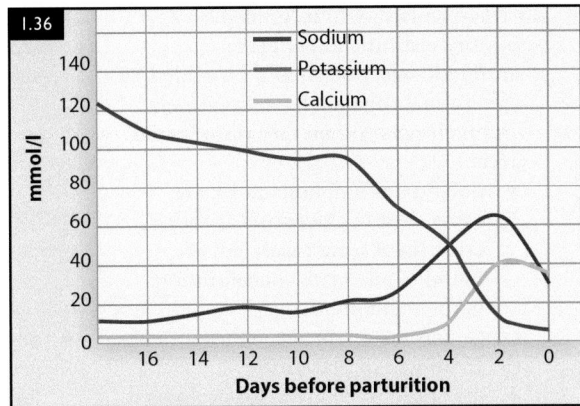

FIG 1.36 Mammary gland electrolyte secretions in the mare.

- ◆ foetus has periods of sleeping and waking.
- ○ assess allantoic fluid:
  - ◆ hyperechoic particles are normally seen.
  - ◆ amount should be monitored over time.
  - ◆ significant increase can suggest foetal/placental compromise.
- ○ caudal pole of the cervix:
  - ◆ placenta attached to uterus with no fluid/pus between the uterus and placenta.
  - ◆ placental separation suggests inflammation/infection.
- ○ combined thickness of the uteroplacental unit (CTUP):
  - ◆ 5–10 cm cranial to the cervical–placental junction and lateral to midline.
  - ◆ identify a uterine blood vessel.
  - ◆ distance between outer uterine wall and inner placental border is measured.
    - – published figures for different breeds are available.
    - – approximate normal thickness for Thoroughbred mare is 1 + gestational age (months).
    - – significant increase suggests placentitis.
    - – significant decrease suggests placental insufficiency.
- • **Transabdominal technique:**
  - ○ placental thickness – caudal pole and uterine horns can be assessed.
  - ○ allantoic and amniotic fluids levels can be assessed and measured.

- ○ foetal presentation can be assessed.
- ○ **foetal measurements:**
  - ◆ orbit size – data for gestational age is available for some breeds.
  - ◆ foetal heart rate – increases during activity.
    - – average heart rate of 76 bpm in the final 2 months of gestation.
    - – persistently elevated heart rate suggests foetal stress.
    - – persistently lower heart rates suggest foetal depression or compromise.
    - – rates are unreliable if the mare is sedated.
  - ◆ aortic diameter – assessment of foetal size.
- ○ foetal breathing movements.

## Management of 'high-risk' mares

- • identification of the high-risk mare can be a challenge to the clinician:
  - ○ often made with the benefit of hindsight and previous breeding problems.
- • any pre-existing maternal illness should be diagnosed, monitored and managed.
- • identification of any placental/foetal compromise involves regular/sequential monitoring:
  - ○ milk electrolytes; progesterone levels; hyperechoic particles in the allantoic fluid.
  - ○ serial examinations allow these changes to be assessed over time.

- any treatments should be guided by laboratory tests and experience:
  - antibiotic selection should be guided by effectiveness against bacteria present and their penetration into the affected area:
    - potentiated sulphonamides are often used in placentitis as they cross the placenta well and are found in adequate concentrations in the foetal fluid.
  - exogenous progestogens are useful in the treatment of placentits:
    - may need to be given for prolonged periods but stopped well ahead of parturition.
- mare should foal under experienced stud management or veterinary supervision:
  - mare may require treatment.
  - foal may be compromised at birth requiring treatment.
    - supplementary oxygen, intravenous fluids, blood analysis including IgG, intravenous plasma, antibiotics, tetanus antitoxin.
- any mare that has had a compromised pregnancy previously should be classed as 'high risk' for subsequent pregnancies and monitored appropriately.

## COMPLICATIONS OF PREGNANCY

## Uterine torsion

### Definition/overview

- uncommon condition occurring in mid–late gestation (rarely at parturition).
- presents as low grade, intermittent abdominal pain.

### Aetiology/pathophysiology

- usually after 7 months of pregnancy.
- cause is unknown but may be associated with foetal activity or mare rolling.
- torsion occurs cranial to the cervix and can be 90–720°.
- interruption of uterine blood/lymphatic system leading to foetal hypoxaemia and placental compromise.

### Clinical presentation

- degree of pain related to degree of torsion and associated vascular compromise:
  - often mild and intermittent abdominal pain 'colic' like.
  - misinterpreted as foaling or aborting.
- complications in severe torsions include concurrent GIT involvement or uterine rupture.
- rarely occurs at parturition leading to dystocia.

### Differential diagnosis

- other causes of abdominal pain e.g., colic, foetal hypermotility.
- abortion and foaling.

### Diagnosis

- rectal palpation:
  - tension and positioning of uterine broad ligament indicates the direction of the torsion.
  - tight band diagonally across the abdomen with the other ligament moving ventrally.
- vaginal examination not usually helpful as torsion is cranial to cervix.
- transabdominal ultrasound to assess foetal health and gastrointestinal disease.

### Treatment

- early intervention essential to prevent foetal/placental compromise.
- non-surgical treatment – under general anaesthesia:
  - wooden plank is used to stabilise the uterus and the mare is 'rolled around' the uterus to correct torsion.
  - increased risk of uterine rupture and is difficult to perform.
- surgical – standing flank laparotomy or ventral midline laparotomy (under GA).

○ latter allows full assessment of the gastrointestinal tract and vascular supply.
• mare should be classified at a 'high risk' and managed appropriately.

## Prognosis

• depends on the degree and duration of the torsion:
  ○ better outcome if torsion occurs prior to 320 days.
• studies suggest overall 56% foal survival and 86% mare survival:
  ○ changes depending on complications and stage of presentation.

# Hydrops amnion/allantois

## Definition/overview

• occurs sporadically in late gestation.
• more common in multiparous mares.
• abnormality of foetal membranes:
  ○ either amnion, allantois or both together.
• large increases in volume of placental fluid leads to acute abdominal enlargement of mare:
  ○ normal placental fluid volume
  ○ allantoic 8–15 l    ○ amnion 3–5 l.
  ○ hydrops allantois cases can have volumes of up to 120–220 l.
• rectal examination and ultrasound confirm diagnosis.
• treatment – induction/abort pregnancy.

## Aetiology/pathophysiology

• Hydrops allantois more common than hydrops amnion.
• placental abnormalities (and/or foetal abnormalities) lead to excessive fluid accumulation.

## Clinical presentation

• late gestation (last trimester).
• rapid fluid accumulation and rapid abdominal enlargement (days/weeks) (Figs. 1.37, 1.38).
• can result in circulatory problems, uterine rupture, and abdominal wall rupture.
• foetal abnormalities are commonly found.

## Diagnosis

• clinical signs are highly suggestive.
• rectal palpation – massively enlarged uterus with often no palpable foetus.
• ultrasonography (transrectal/transabdominal) reveals extensive fluid accumulation:
  ○ may be able to identify which fluid is involved (amnion/allantois).

## Management

• mare is unlikely to be able to sustain the pregnancy without significant complications.
• very high likelihood of foetal abnormality and non-viable foetus.
• termination of pregnancy – abortion or induction:

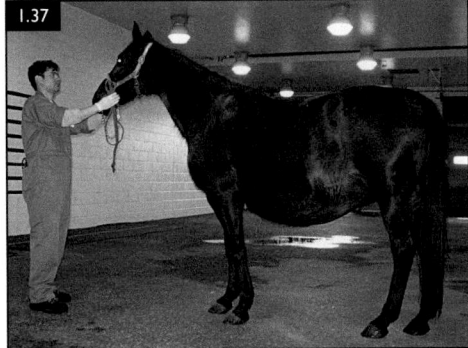
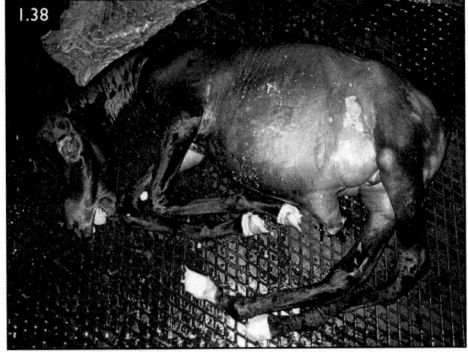

**FIGS. 1.37, 1.38** (1.37) A late-pregnant mare showing gross abdominal enlargement due to a hydrops. (1.38) After parturition the foal was found to be abnormal, with a hydrocephalus and grossly enlarged abdomen.

- assisted delivery due to high likelihood of uterine inertia.
- controlled removal of placental fluid is essential:
  - circulatory shock and electrolyte imbalances are risks.
  - sterile nasogastric tube used to siphon the fluids off at a controlled and steady rate.
  - intravenous fluid therapy (including hypertonic saline) at time of drainage to maintain blood pressure.
- placenta is frequently retained and should be treated as soon as possible.
  - oxytocin drip stimulates uterine contractions and prevents uterine pooling.

## Prognosis

- advanced or chronic cases:
  - ventral wall rupture and circulatory compromise can occur – poor prognosis.
- foal is usually abnormal/non-viable so prompt termination is advised for mare's welfare.
- prompt treatment aims to preserve the mare's future breeding potential.
- subsequent fertility is not usually affected.

# Rupture of the ventral abdominal muscles/ prepubic tendon

## Definition/overview

- usually older, multiparous mare with previous damage/stretching of supporting structures.
- initially presents with ventral oedema and mild abdominal pain before abdomen drops.
- ultrasonography of the abdominal wall may show rupture.
- treatment includes an abdominal support, box rest, and assistance at foaling.
- future breeding is compromised due to rupture.

## Aetiology/pathophysiology

- progressive rupturing of the prepubic tendon and/or rectus/transverses abdominus muscles occurs in late gestation.

- strain from previous pregnancies can cause weakening of the structures.
- twin pregnancy, hydrops, multiparity, increasing age, and draught breeds are all considered predisposing factors.

## Clinical presentation

- ventral abdominal oedema (up to 15 cm) coursing from in front of the udder to the front legs (Fig. 1.39, 1.40).
- can be painful with mare unwilling/ unable to walk.
- dropping of the abdomen or asymmetric 'bulging' indicates muscle wall rupture.
- loss of usual abdominal contour and cranial movement of the udder:
  - indicates prepubic tendon rupture.
- pelvis tilts forward (tail head raised) as the weight distribution alters (lordotic stance).
- udder secretions are often blood tinged due to the trauma.

## Differential diagnosis

- other rupture/herniations including post-surgical and traumatic causes.
- oedema from mild vascular compromise is often seen in late-gestation mares:
  - must be differentiated from more serious ruptures.

## Diagnosis

- clinical history and appearance.
- transabdominal ultrasound (Figs. 1.41–1.44).
- rectal examination and ultrasound may identify prepubic tendon damage cranial to pubis.

## Management

- exercise restriction (large stable/foaling box).
- abdominal support bandage (Fig. 1.45):
  - can be difficult to manage and cause skin trauma.
- pain relief – usually NSAIDs.
- laxatives.
- severe or deteriorating cases:
  - induction of parturition should be considered.
  - only after checking foetal readiness for birth (milk electrolytes).

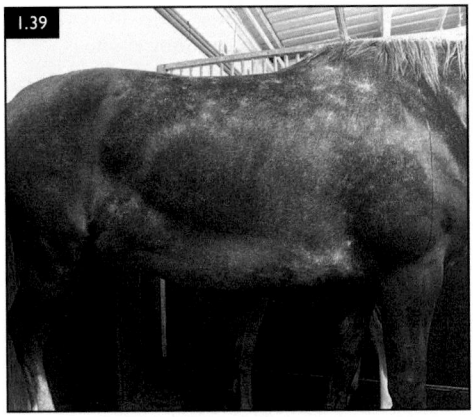
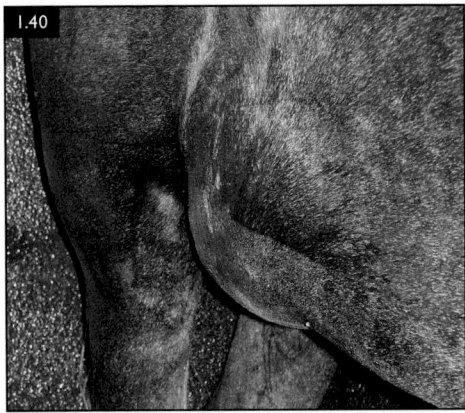

**FIGS. 1.39, 1.40** (1.39) Large Danish Warmblood mare on the day of a normal foaling. In the weeks prior to foaling there had been increasing ventral oedema and stiffness in the mare. (1.40) Immediately post foaling, an additional swelling had occurred on the lower right side of the abdomen.

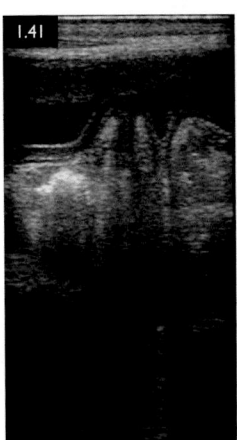
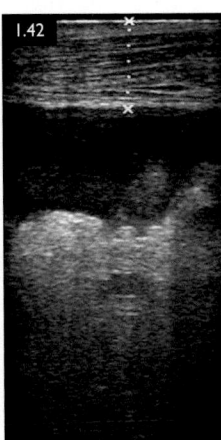
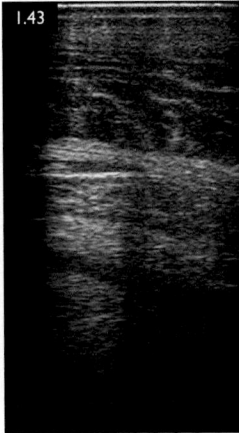
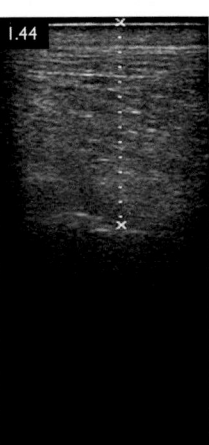

**FIGS. 1.41–1.44** Ultrasonograms of the horse in 1.39 and 1.40. (1.41) Directly over the right-sided swelling showing complete loss of the abdominal wall at the right side due to rupture; (1.42) the left side showing normal wall thickness; (1.43) over the ventral midline revealing extensive subcutaneous oedema over a thickened abdominal wall; (1.44) over the prepubic tendon, just cranial to the pubis, showing thickening and disruption of the fibre pattern of the tendon.

- o foetal maturation can be attempted by treating the mare with dexamethasone (100 mg per 500 kg horse i/m q24h for 3–5 days) where the mare is severely compromised.
- foaling should be assisted as mare's abdominal efforts will be reduced.

## Prognosis

- damage to the soft tissues can be variable but may be so severe that mare euthanasia on humane grounds is recommended.
- prepubic tendon damage cannot be repaired.

- abdominal muscle rupture usually does not heal sufficiently to allow future breeding:
  - o mesh repair can be considered but is usually unsuccessful due to extent of damage.
- if the mare is comfortable and it is recommended that she does not breed again:
  - o assisted reproductive techniques could be considered.
    - ◆ embryo transfer
    - ◆ oocyte pickup/ICSI.

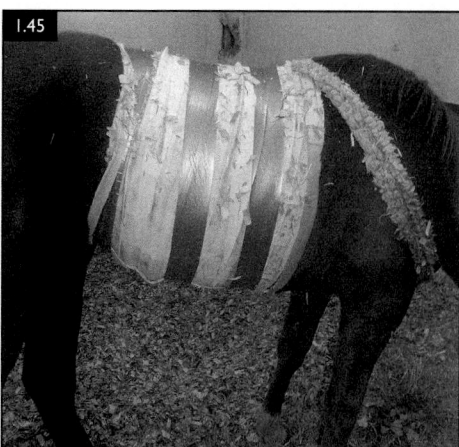

FIG. 1.45 Use of an abdominal support bandage in a mare after foaling that had early signs of ventral rupture of the abdominal wall prior to parturition.

## Twins

### Definition/overview

- common in some breeds e.g. Thoroughbred (20–30%) but rare in ponies.
- occurs when multiple ovulations (synchronous or asynchronous) are fertilised resulting in two or more embryos:
  ○ identical twins, from the splitting of a fertilised ovum, is extremely rare in the horse.
- horse is unable to sustain a twin pregnancy to term except in rare cases (Fig. 1.46):
  ○ such cases often have significant complications.
- twins that implant in the same horn are termed *unilateral*.
- twins that implant in opposite horns are termed *bilateral*.
- ideally the number of ovulations will be known for all mares from pre-covering scanning.
- **twins/multiple pregnancies must be treated and monitored to ensure that only a single pregnancy remains.**
- early recognition by scanning is essential:
  ○ ideally when conceptuses are still mobile in pre-implantation period (<16 d after ovulation).

- ○ reduces the likelihood of damage to the remaining pregnancy if one is destroyed.
- most twin pregnancies fail by aborting in mid–late gestation.

### Aetiology/pathophysiology

- multiple ovulations.
- unilateral twins are competing for the same endometrial contact for the placenta:
  ○ generally unsustainable resulting in loss of one or both pregnancies.
- bilateral twins can develop until mid-pregnancy:
  ○ start to compete with each other for space, often resulting in abortion.
- rarely carried to term but if they are they are usually small/dysmature:
  ○ associated with dystocia, abdominal wall/prepubic tendon rupture.
- rarely one foal dies and is mummified with the remaining twin going on to develop:
  ○ foal born with the mummified foetus being discovered at parturition.

### Clinical presentation

- pre-covering examination will identify two or more pre-ovulatory follicles:
  ○ use of ovulatory agents increases the likelihood of multiple ovulations.
- twin conceptuses/foetuses by rectal palpation or ultrasonography during pregnancy:

FIG. 1.46 A remarkably uncommon situation in which twin foals have been born live to a pony mare and survived. Both were dysmature at birth. Note the disparity in the size of the two foals.

- up to 14-18 days post ovulation the twin conceptuses may be mobile.
- abortion of twins.
- abnormal abdominal enlargement of the mare.
- abdominal wall or prepubic tendon rupture and ventral oedema.
- premature udder development.

## Diagnosis

- ultrasound examination of the entire reproductive tract 14–15 days post ovulation:
  - identify the still mobile embryos (Fig. 1.47).
  - include an ovarian examination to count the number of corpora lutea present.
  - mares with a history of 2+ ovulations should be checked again 48–72 hours later to confirm no twin pregnancies.
  - asynchronous ovulations may result in different sized embryos that are missed if only one examination is carried out:
    - ♦ repeat examinations are warranted.
  - Endometrial cysts can appear similar to the early conceptus:
    - ♦ mapping their size, shape and location pre-breeding will help distinguish them from a pregnancy.
    - ♦ re-examinations are necessary to confirm that the cyst is not an embryo:
      - – endometrial cysts do not grow, change shape or move around the uterus.

## Management

- multiple techniques available depending on the timing of presentation:
  - essential that mare is examined, treated and monitored until the clinician is confident that no twin pregnancy is present.
    - ♦ usually requires multiple examinations.
  - whole uterus must be examined fully requiring the mare to be:
    - ♦ compliant, in stocks, sedated, and with subdued lighting etc.
    - ♦ referral to an experienced colleague may be required in some cases.

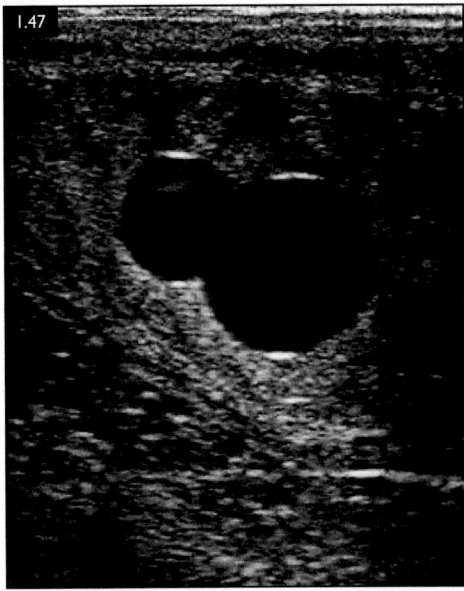

**FIG 1.47** Twin pregnancy. Two pregnancies (13 days and 14 days) adjacent to each other in the same uterine horn.

- **Pre-fixation (<16 days post ovulation):**
  - best time for examinations as the embryos are mobile and small.
  - smallest vesicle is identified and carefully moved away from the other vesicle:
    - ♦ use the rectal probe avoiding traumatising the uterus or larger vesicle.
  - small vesicle is moved to the tip of the uterine horn.
  - gentle pressure is applied to 'squeeze' the vesicle.
    - ♦ characteristic slight 'pop' may be felt.
    - ♦ free fluid may be seen where the vesicle was previously.
  - check the remaining vesicle is still present.
- **Post-fixation (17–30 days):**
  - bilateral twins can be reduced as described above:
    - ♦ pregnancy is disrupted but often remains visible (especially in 20+ days).
  - unilateral twins are difficult as there is a high risk of traumatising both twins:

- ♦ if the vesicles are small repeated gentle pressure can be applied on repeated examinations to produce a 'snowflake' appearance (see below).
    - – excessive trauma should be avoided.
  - ○ twins resulting from asynchronous ovulations will have a size difference:
    - ♦ difference is 4 mm or greater – high chance the smaller unilateral twin will be reduced by the mare.
    - ♦ repeated examinations are necessary to ensure this occurs otherwise PG (see below may be given)
- **Prostaglandin** can be given up until day 35 (prior to eCG production):
  - ○ pregnancy is terminated and the mare will return to oestrus and can be re-bred.
  - ○ used in cases of unilateral twins that cannot be reduced.
- **Post-fixation (30–45 days):**
  - ○ bilateral twins should be pinched, or pressure applied, via the rectal probe to create the 'snowflake' effect indicting cellular debris and damage within the pregnancy.
  - ○ pregnancy should not be ruptured as the fluid can compromise the remaining twin.
  - ○ other methods include:
    - ♦ transvaginal ultrasound-guided twin puncture:
      - – specialised technique to aspirate fluid from the pregnancy causing its demise.
      - – ideally performed prior to 45 days.
    - ♦ transabdominal ultrasound-guided twin puncture:
      - – specialist technique to inject potassium chloride or procaine penicillin into the twin (100 days onwards).
    - ♦ craniocervical dislocation:
      - – specialist technique to dislocate/ decapitate a twin via transrectal manual manipulation or more commonly a standing flank laparotomy.
- all techniques require monitoring to ensure a single pregnancy remains and the treated twin regresses.

- some clinicians use flunixin meglumine at the time of early twin reduction.
- Altrenogest can be used together with flunixin meglumine in the 17–45 days twin reductions to help maintain the remaining twin.
- Flunixin meglumine, altrenogest, NSAIDs and antibiotics are used in the 'other methods' due to their invasive nature.

## Prognosis

- manual reduction of twins, by experienced clinicians, is very successful in the pre-fixation period with no increase in pregnancy loss compared to the normal mare pregnancy.
- later twin reduction techniques are less successful (50%) and significant for the mare.
- leaving a twin pregnancy should be avoided at all costs:
  - ○ carrying twins to later gestation can lead to major complications.
  - ○ impact future mare fertility.

# Placentitis

## Definition/overview

- common cause of sporadic abortion and vaginal discharge in late pregnancy.
- ascending infection via the cervix is most typical although haematogenous spread can occur.
- bacterial (most common), fungal and viral causes have been identified.
- results in:
  - ○ generalised placentitis
  - ○ premature placental separation.
  - ○ foetal compromise/growth retardation/ abnormal maturation/death.
  - ○ abortion, stillbirth or mummification.
- early diagnosis is difficult but essential if treatment is to be successful.
- live foals may be delivered but considered high risk for development of perinatal asphyxia syndrome and septicaemia.

## Aetiology/pathophysiology

- typically ascending bacterial infection through the cervix:
  - ○ commonly because of poor vulva/ vestibule seal and pneumovagina.

- ◆ older, multiparous mares with poor perineal conformation may have an increased risk of placentitis.
  - ○ less commonly iatrogenic, following contact with infected fomites or at breeding.
  - ○ haematogenous spread is possible, particularly *Nocardia* placentitis (USA):
    - ◆ causes lesions at the base of the uterine horns.
- **Bacterial** causes include:
  - ○ *Streptococcus* spp. (most common isolate), *Staphylococcus* spp., *Escherichia coli, Pseudomonas* spp., *Klebsiella* spp., *Salmonella abortus equi, Corynebacterium pseudo-tuberculosis, Leptospira pomona* (haematogenous), *and Nocardia* spp.
- **Viral** causes:
  - ○ EHV-1 (most commonly) or EHV-4
  - ○ EVA virus.
- **Fungal** causes are rare and most are caused by *Aspergillus* spp.
- infection leads to:
  - ○ chorionic villi inflammation and necrosis.
  - ○ interference with chorion/endometrial interdigitation.
  - ○ placental insufficiency.
  - ○ thickens the placenta, often at the caudal pole.
  - ○ often leads to premature placental separation and either premature delivery or foetal death.
- infection can spread from the placenta directly to the foal resulting in foetal death from septicaemia.
- infected foetal membranes and fluids, as well as uterine secretions from aborted mares:
  - ○ source of infection in the viral (EHV and EVA) causes.
  - ○ latent infections are possible with EHV-1, which can be activated by stress.
- placentitis affecting only a limited area of placenta, or a severe placentitis occurring close to full term:
  - ○ may not cause abortion but may result in delivery of a live foal.
  - ○ foal is often weak, poorly grown, dysmature, or septicaemic.
- foetal stress can occur in late term placentitis and cause sufficient early

maturity of the foetal adrenal cortex to occur, which allows the foal to survive despite its early gestational age.

## Clinical presentation

- premature mammary gland development and lactation.
- depending on the cause, other signs may include vulval discharge, pyrexia, depression, and anorexia.
  - ○ vulval discharge is variable and may be difficult to identify:
    - ◆ check tail/perineum for dried discharge (Fig. 1.48).
  - ○ viral placentitis can present with respiratory signs and peripheral oedema.
  - ○ abortion/premature delivery may be immediate/delayed, and often sporadic (Fig. 1.49).
    - ◆ EHV placentitis can result in multiple abortions in a herd (abortion storms).
  - ○ any post-partum discharge should be investigated as potential sign of a pre-existing placentitis and the foal monitored as high-risk.

## Differential diagnosis

- other causes of abortion.
- other causes of vaginal discharge:

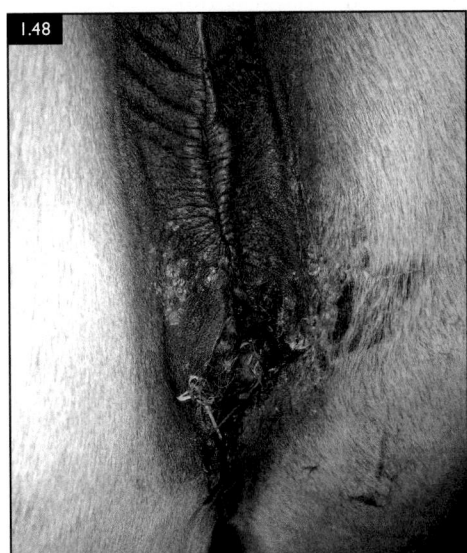

**FIG. 1.48** Vulval discharge in a pregnant mare with placentitis.

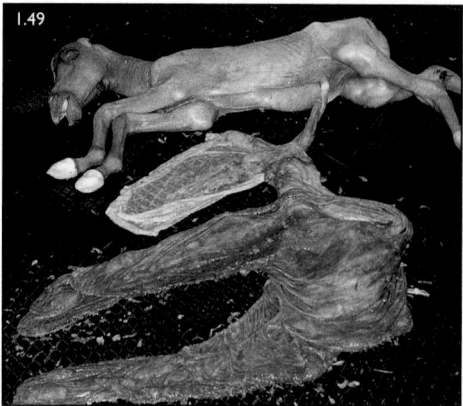

**FIG. 1.49** Aborted, poorly grown foetus subsequent to a placentitis.

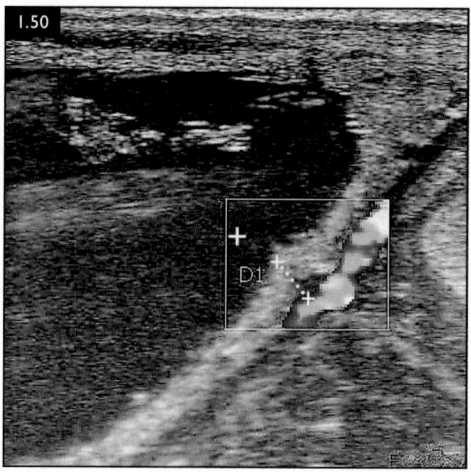

**FIG. 1.50** Ultrasound image allowing measurement of the CTUP. The distance between the uterine vessel and allantoic fluid is measured three times (x-x on screen) and averaged. This is the CTUP measurement.

- urine staining (urogenital disorders)
- vaginal varicose veins.
- trauma to caudal reproductive tract.

## Diagnosis

- any vaginal discharge sampled for culture and cytology.
- transrectal ultrasonography to assess:
  - foetal movement
  - caudal placental thickness
  - placental appearance.
  - placental fluid appearance
  - placental separation.
  - CTUP measurement (Fig. 1.50).
    - ♦ transrectal ultrasound with probe placed just cranial to the cervix.
    - ♦ move laterally to image a uterine blood vessel on ventral border of the uterus.
    - ♦ measure the distance in three areas and take average.
    - ♦ can be repeated to monitor response to treatment.
- transabdominal assessment of foetus.
- examine the placenta after abortion/ parturition for any gross lesions (Fig. 1.51):
  - samples should be taken for culture and histopathology.
- blood sample analysis:
  - inflammatory markers are particularly sensitive for diagnosis and monitoring response to treatment.
  - progesterone and oestrogen assays are helpful in diagnosis.

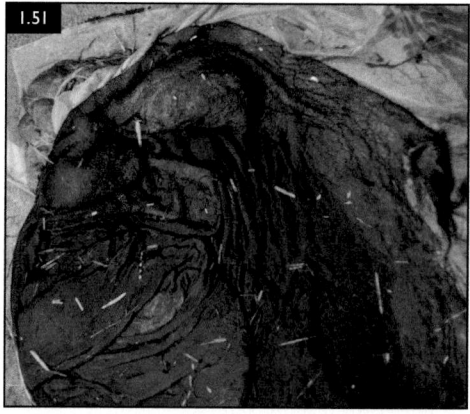

**FIG. 1.51** Part of the placenta of a Thoroughbred mare that had foaled following a late-pregnancy vaginal discharge. There are areas of placental thickening and brown discolouration due to a focal placentitis of bacterial origin.

## Management

- treatment is aimed at:
  - controlling infection, reducing inflammation and encouraging uterine quiescence.
- Antibiotics:
  - broad-spectrum or guided by culture/ sensitivity.
  - good penetration of foetal and placental tissue.

- potentiated sulphonamides
  - penicillin/gentamicin.
- Anti-inflammatories:
  - firocoxib, flunixin, phenylbutazone and meloxicam.
- Progesterone supplementation eg altrenogest (0.088 mg/kg q24h p/o).
- Pentoxyifylline.
- treatment is continued until the foal is delivered or if placentitis is identified early:
  - monitor response to treatment to guide antibiotic duration.
  - continue altrenogest treatment for entire pregnancy.
- any foal delivered from a placentitis mare should be classed as high-risk.
- prevention should involve:
  - strict stud hygiene
    - immediate isolation of aborted mares.
    - remove all contaminated placental fluid/placenta/foetus after parturition and abortion.
    - isolate mare and in-contacts until clearance if infectious cause is suspected.
  - use of EHV and EVA vaccines where available.
  - minimise stress to pregnant mares.
  - vulval conformation problems should be addressed before future pregnancies.

## Prognosis

- early identification and aggressive treatment – up to 73% of cases may result in a live foal.
- inflammation is often well advanced when identified and abortion is a common outcome.

## Foetal mummification

### Definition/overview

- uncommon.
- occurs if intrauterine foetal death occurs and the cervix remains closed:
  - foetus is retained and may become 'mummified'.
- occur in twin pregnancy if one foetus dies and the other continues to develop to term.

- single foal cases, which are rare, the mare may spontaneously abort the dead foetus at any time or enter an extended period of anoestrus.

## Aetiology/pathophysiology

- placentitis may cause foetal death and treatment (particularly altrenogest) may keep the cervix closed leading to retention of a dead foetus.
- twin pregnancy is the most likely scenario to result in mummification (Fig. 1.52):
  - one of the foetuses spontaneously dies and subsequently mummifies, while the pregnancy is maintained by the remaining live foetus.
  - if foetal tissue remains sterile, abortion/evacuation may not occur or be delayed.
- mummified foetus may macerate *in utero* over time or as a result of an ascending infection via the cervix, leading to a vulval discharge.

## Clinical presentation

- spontaneous abortion at any time in pregnancy but commonly second trimester.
- prolonged anoestrous or variable cyclicity when mare assumed in foal.
- rarely mummified foetus delivered singly or alongside normal foal at parturition.
- foetal maceration – brown vulval discharge may be seen.

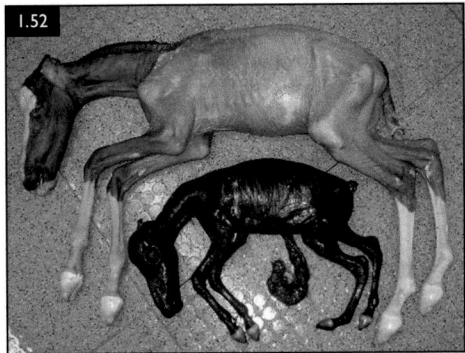

**FIG. 1.52** Twin abortion in the last trimester of pregnancy showing clear disparity in foetal size. The smaller foetus had already died *in utero* and undergone early mummification.

## Differential diagnosis

- other causes of abortion.

## Diagnosis

- clinical presentation and history.
- rectal palpation reveals enlarged uterus contracted around a dry, contorted foetus with no fluids.
- ultrasonography shows an abnormal foetus or uterine contents with little uterine fluid.

## Management

- induction of parturition/abortion in the late-term mare:
  - prostaglandin/oxytocin/cervical dilation.
  - lubrication may be necessary to aid passage of the foetus.
- culture/sensitivity and histopathology of foetal remnants.
- uterus should be flushed with sterile saline post-partum to remove debris and contamination.
- cases of maceration, where infection is suspected, antibiotics should be given based on culture and sensitivity results.
- caesarean section is rarely required to remove a large, mummified foetus.

## Prognosis

- undiagnosed cases may result in loss of breeding season due to anoestrus/uterine infection.
- no long-term fertility problems should be expected unless uterine trauma/infection occurs.

# Early embryonic death (EED)

## Definition/overview

- death of embryo before day 40 of pregnancy.
- most occur prior to first pregnancy diagnosis (14–15 days post ovulation).
- number of proposed causes.

## Aetiology/pathophysiology

- **Foetal factors:**
  - embryonic/chromosomal/ developmental abnormalities.
  - inbreeding.
- **Maternal factors:**
  - aged oocyte/sperm at conception.
  - individual mare or stallion genetic factors.
- **Management factors:**
  - stress (hyper/hypothermia, transport)
  - poor nutrition.
  - concurrent systemic illness.
  - chronic or acute pain
  - toxicosis   o drug therapy.
- **Individual mare factors:**
  - age   o uterine infection
  - breeding at foal heat.
- **Uterine factors:**
  - delayed uterine involution post-partum (foal-heat breeding).
  - endometritis.
  - chronic endometrial disease (uterine atrophy, fibrosis).
  - endometrial cysts.
- **Endocrine factors:**
  - inadequate progestogen production by primary CL.
- **Others:**
  - twin pregnancy.
  - prostaglandin administration

## Clinical presentation

- mare returns to oestrus after positive pregnancy diagnosis.
- mare not in foal when examined after a positive pregnancy diagnosis.

## Differential diagnosis

- see pathophysiology section.
- Oviductal obstruction

## Diagnosis

- regular ultrasound examinations demonstrate loss of conceptus/growth rate abnormality:
  - conceptus should grow at a predictable rate.
  - any deviation from this suggests a compromised pregnancy.

## Management

- close monitoring of pregnancy to see if it will 'catch up' to expected appearance.
- attempts made to 'rescue' the pregnancy with exogenous progestagens (eg altrenogest):

- o unlikely to be successful and just delays the pregnancy failure.
- prostaglandin to induce luteolysis and return to oestrus:
  - o investigate and treat any uterine abnormality and re-breed.
- mares prone to EED can be:
  - o treated with exogenous progestagens to support the primary CL.
  - o progestogen assay performed at 5–7 days after ovulation to assess CL function.
- Buserelin at day 10 post ovulation reported to improve pregnancy rates in some cases.
- uterine biopsy of the mare to assess endometrial health.
- oviductal treatment with prostaglandin E via laparoscopic or hysteroscopic application:
  - o encourage flushing of debris from the oviducts.
  - o considered in older mares for which there are no other causes of infertility or EED.

## Prognosis

- good to poor depending on the cause.

# Gastrointestinal complications of late pregnancy and parturition

## Definition/overview

- late-term mare is at risk due to large uterus in abdomen.
- during foaling due to foal movements:
  - o caecal or right ventral colon trauma/ rupture.
  - o contusion of small intestine/small colon and attaching mesentery:
    - ◆ damage can lead to secondary rupture.
- uterine damage also often occurs.
- colonic torsion is an increased risk within the first 4 weeks post-partum (Fig. 1.53).

## Aetiology/pathophysiology

- caecum/large colon can be traumatised during parturition, particularly if full of ingesta/gas.
- prolonged straining, foetal limbs, and bowel trapped between the pelvis and

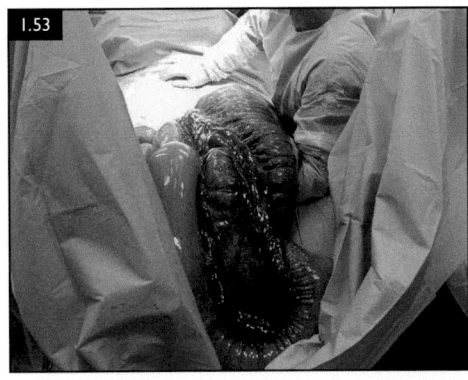

**FIG. 1.53** Surgical intervention in a broodmare 7 days post foaling that revealed a 360° large colon torsion. Note the enlarged and discoloured left dorsal and ventral colon, with a haemorrhagic mesentery.

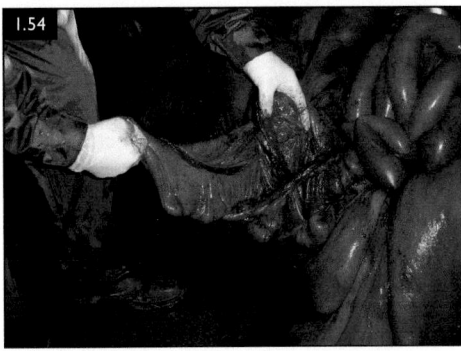

**FIG. 1.54** Post-mortem view of a broodmare that was euthanased after a post-foaling peritonitis. Note the rupture in the caecum.

foetus can all lead to gastrointestinal damage.
- normal increased abdominal pressure during parturition can lead to:
  - o spontaneous rupture of the large colon/ caecum.
    - ◆ primary site within 15 cm of the ileocaecal junction.
    - ◆ ventrally in caecum and caudally in right ventral colon (Fig. 1.54).
- contusions of small intestine, small colon, rectum and mesentery can occur at parturition:
  - o severe damage can lead to rupture, incarceration, vascular compromise, and segmental ischaemic necrosis.
- external prolapse of small intestine/small colon can occur through the rupture leading to gross contamination.

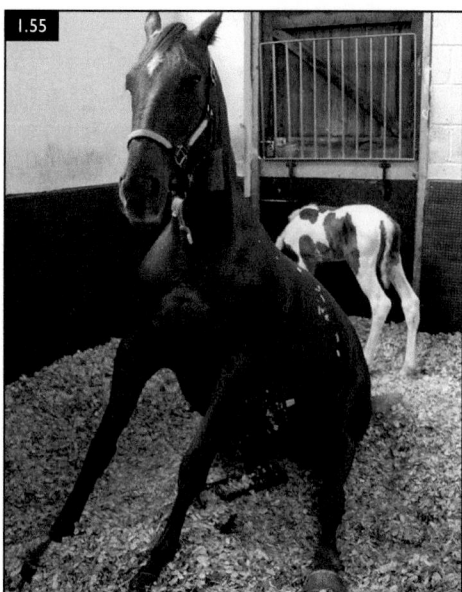

FIG. 1.55 'Dog-sitting' behaviour in a broodmare with abdominal pain a few days post foaling.

## Clinical presentation

- related to the degree and site of damage (Fig. 1.55).
- rupture of the caecum/colon early in parturition:
  - mare may present with dystocia or weak abdominal straining.
  - foal can usually be delivered with assistance.
  - mare does not recover normally from foaling:
    - ◆ rapidly develops signs of septic peritonitis and endotoxic shock.
    - ◆ intestinal prolapse can also occur.
    - ◆ death can occur within 4–6 hours.
- less severe trauma can lead to slower onset of signs over several days including:

  - delivery may or may not have been complicated.
  - milk production is reduced and the foal can be seen to persistently nurse.
  - colic, pyrexia, depression and peritonitis.
  - bowel damage can lead leakage of bacteria, endotoxaemia and death.

## Diagnosis

- rectal palpation can have variable findings:
  - normal.
  - impacted small colon
  - painful foci in the abdomen.
  - roughened, inflamed serosal surfaces.
- peritoneal fluid collection and analysis:
  - colour and smell may indicate rupture or iatrogenic bowel puncture.
  - cytological analysis for peritonitis diagnosis.
- transrectal/transabdominal ultrasonography and laparoscopy may be helpful.

## Management

- severe rupture with systemic compromise
  - euthanasia is indicated.
- surgical laparotomy or laparoscopic assessment and repair is possible if identified early but surgical access can be limiting and lesions inoperable.
- peritonitis treatment:
  - antibiotics, NSAIDS, intravenous fluids, intensive care and peritoneal lavage/drainage.

## Prognosis

- poor to grave depending on degree, site and extent of damage and peritoneal contamination.

## ABORTION

## Definition/overview

- expulsion of foetus and placenta prior to 300 days of gestation (Figs. 1.56–1.58).
- variable incidence – reported as 5–15%.
- causes can be divided into infectious, non-infectious and unknown:

  - non-infectious causes are the most common (up to 70%).
  - no cause is identified in many cases (up to 50%).
- often sporadic, isolated occurrences but cases of infection can cause multiple abortions ('abortion storms').

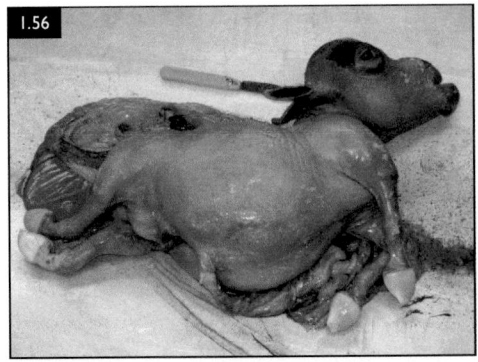

**FIGS 1.56–1.58** (1.56) Fresh aborted foetus with multiple abnormalities, including the mandible, and small size. (1.57) Foetal abortion due to excessive umbilical cord length and torsion causing umbilical vessel thrombi to develop, leading to foetal death and abortion. (1.58) A grossly normal foetal abortion must still be investigated fully, but often no cause is identified.

## Clinical presentation

- varies depending on the cause and timing of abortion:
  - some mares found to be non-pregnant at end of gestational period:
    - without any foetal loss being detected.
  - premonitory signs are uncommon but mares with late gestation placentitis can develop:
    - early mammary gland enlargement, lactation and vulval discharge.
- actual abortion is often unseen:
  - if seen, the mare will present with signs of parturition.
- foetus is frequently discovered in the stable or field after the event:
  - identifying the affected mare may be difficult if the latter occurs:
    - check vulva or tail for discharge.
    - rectal palpation/ultrasound may be necessary.
    - occasionally the foetus and placenta may be scavenged by wild mammals and go unnoticed.
- Dystocia is rare.

## Management

- essential to approach abortion thoroughly to allow prompt identification of the cause (if possible) and minimise risk to other mares on the stud or wider population.
- approach includes:
  - isolate the mare and the foetal/placental remains.
  - keep in-contacts separate and under observation until the cause is ascertained.
  - examine the mare, foetus and placenta thoroughly:
    - rarely complications in the mare as complete foetoplacental unit is often delivered.
    - submit post-mortem samples for histopathology (formalin fixed) and EHV PCR (plain) to a competent laboratory.
    - foetal samples to include:
      - thymus, lung, spleen, liver, kidney.
      - look for excessive/discoloured abdominal fluid or perirenal fluid.
    - placental samples to include:
      - cervical pole, pregnant and non-pregnant horn.
    - include history of management, vaccination status, gross post-mortem findings.
  - alternatively complete foetus and placenta submitted to the laboratory for post-mortem.

- every abortion should be treated as potentially infectious until proven otherwise:
  - biosecurity measures remain in place until negative laboratory results are available.
- management of abortion, including guidance on EHV abortion management, should refer to the country or region-specific disease control programmes e.g. Horserace Betting Levy Board's Code of Practice in UK https://codes.hblb.org.uk or EquiBioSafe app.
- complications following an abortion are uncommon if all material is expelled:
  - mare should be examined internally together with the aborted material to confirm no parts of the placenta has been retained.
- after any necessary isolation the mare should be returned to normal management and re-establish normal reproductive function.

## Prevention of abortion:

- identify any endometrial disease that may lead to late-term foetal compromise.
- appropriate early pregnancy ultrasound examination to confirm a single developing foetus with normal growth.
- identify and treat any vulval conformation problems and correct e.g. Caslick's procedure.
- identify and treat any cervical abnormalities prior to covering.
- identify other causes of ascending placentitis and treat e.g. exogenous progestagens.
- vaccinate for EHV1/4 and EVA:
  - cases of EHV abortion, vaccination extended to other horses on the farm:
    - particularly weanlings, yearlings and competing/racing horses.

# Infectious causes of abortion

# Equine Herpesvirus type 1

## Definition/overview

- most important cause of infectious abortion.
- sporadic cases or extensive numbers 'abortion storms'.

- respiratory and neurological forms also exist.
- whole horse population approach needed to reduced incidence of disease.
- outbreaks can have significant financial impact on breeding farms.

## Aetiology/pathophysiology

- virus transmitted by aerosol via respiratory route.
- spreads slowly between in-contact horses.
- respiratory infection in non-pregnant horses is the major risk to pregnant mares.
- material from EHV abortion (foetus, placenta, foetal fluid) is highly infectious.
- suggested that a vasculitis of blood vessels within the endometrium and/or chorion due to antibody/antigen complexes leads to rapid abortion.
- carrier status can occur.

## Clinical presentation

- most abortions occur after 5 months of gestation (around 9 months) and 4–14 weeks after infection.
- foal is usually aborted within the placenta with no premonitory signs.
- usually no autolysis of the foetus.
- occasionally the aborted foal is born alive but dies within 24 hours.
- mare infected late in gestation may produce a live foal:
  - born with severe pneumonia requiring intensive care/management.
  - condition is often fatal to foal.

## Diagnosis

- based on history, foetal and placental histopathology, and PCR/virus isolation.
- aborted foetus can have characteristic lesions:
  - excessive serosal fluids.
  - small (1 mm) white dots on liver (necrosis).
  - enlarged spleen.
  - pneumonic lung lesions/respiratory exudate
  - pleuritis.
- placental changes can include:
  - oedematous chorion.
  - foal often born in the placenta (chorion on outside).

- histological lesions:
  - foci of necrosis in liver, lungs and spleen.
  - inclusion bodies within the hepatocytes and other cell types.
- PCR testing can be done rapidly on fresh material (not fixed).
- serological changes can be helpful but caution in interpretation, especially in vaccinated mares.

## Management

- no treatment of the aborting mare or aborted foetus.
- mare develops natural immunity following recovery, which lasts for 4-6 months.
  - repeat abortions can occur in consecutive pregnancies.
- vaccination of pregnant mares with a killed vaccine at 5, 7 and 9 months of gestation can decrease the incidence in groups of mares.
- biosecurity and quarantine measures:
  - batching together mares of similar gestational age.
  - quarantine new arrivals.
  - keep stress to a minimum:
    - ♦ avoid mixing mares into new groups
    - ♦ avoid transport.
- keep pregnant mares away from young stock/competing horses (respiratory EHV).

# Equine Herpesvirus Type 2

## Definition/overview

- generally sporadic cases of abortion.
- See EHV1 for further detail and management.

# Equine Viral Arteritis

## Definition/overview

- occurs worldwide in all populations of horses.
- some countries have strict governmental controls to prevent entry of the disease:
  - UK notifiable disease.

## Aetiology/pathophysiology

- EVA can spread via two routes:
  - aerosol contact with the respiratory form of the disease.
  - venereal transmission from:
    - ♦ carrier stallion via natural mating or AI.
- aborted foetus, placenta and foetal fluids are infectious.
- incubation period is 3–8 days, with the disease leading to a generalised vasculitis.
- high morbidity, low mortality.

## Clinical presentation

- variation in clinical signs and severity and include:
  - pyrexia   ○ depression   ○ anorexia.
  - nasal discharge   ○ limb oedema.
  - ventral abdominal oedema.
  - conjunctival oedema ('pink eye').
  - skin plaques.
  - abortion.

## Diagnosis

- challenging prior to abortion.
- paired serology testing to confirm:
  - exposure of the mare to virus and identify carrier stallions.
  - pre-breeding serology testing is mandatory/recommended in some countries.
- virus isolation from aborted foetus, nasal discharge of mare, and semen in carrier stallions.
- vaccination is available for stallions but will cause false positives on serology:
  - specific rules regarding pre-vaccination blood testing and documentation.

## Management

- supportive treatment with most animals recovering.
- prevention includes:
  - pre-breeding testing of all mares and stallions.
  - vaccination if appropriate in stallions.
  - use of carrier stallions should be avoided if possible and treatment considered.
- when an outbreak occurs follow the governmental rules for the specific country e.g. UK https://codes.hblb.org.uk or EquiBioSafe app.

# Bacterial abortion

## Definition/overview

- numerous bacteria can cause abortion in mares.
- particular attention should be paid to reduce the risk factors allowing bacterial invasion.
- monitoring mid–late-term mares help to identify cases early, allowing prompt investigation and treatment.

## Aetiology/pathophysiology

- bacteria can be introduced:
  - at breeding
  - ascend through the vulva/cervix.
  - pneumovagina or cervical incompetence can predispose to ascending infections.
  - spread haematogenously.
- placentitis develops causing placental thickening and exudate formation.
- infection can spread to the foetus.
- abortion occurs following foetal death from direct infection/septicaemia, progressive placentitis or placental insufficiency.
- bacterial agents include:
  - *Streptococcus* spp. (common), *E. coli*, *Pseudomonas* spp., *Klebsiella* spp., and *Staphylococcus* spp.
  - haematogenously spread organisms include *Salmonella* spp., *Streptococcus zooepidemicus*, and *Leptospira* spp.
  - non-cervical placentitis (chorion body and base of the horns) associated with a nocardioform organism, has been reported in the USA, particularly in Kentucky.

## Clinical presentation

- vaginal discharge (variable) and premature mammary gland development/lactation:
  - presentation pre-abortion – mare and pregnancy should be considered high risk:
    - detailed foetoplacental assessment is warranted.
- placental appearance following abortion can vary:
  - minimal in acute cases.
  - thickened, discoloured (often brown) and covered in exudate.
  - either generally or at the cervical pole.

## Diagnosis

- samples from placenta, foetus or uterus can identify bacterial infection:
  - culture/sensitivity and cytology.
  - Leptospiral abortion:
    - difficult to diagnose as hard to culture and present with variable findings.
- immunofluorescence and specialised histopathological techniques (staining) may identify causative organisms.
- rising antibody titres can be helpful in some cases.

## Management

- treatment and prevention of placentitis is covered earlier.
- vaccination/off-licence vaccination for some causes may be possible in some countries.
- special attention to management and hygiene to prevent other mares being exposed.
- prognosis is better for cases identified earlier but is generally guarded to poor.

# Fungal abortion

## Definition/overview

- rare cause of late abortion (> 10 months), stillbirth or congenitally infected live foal.
- usually, an ascending infection due to cervical or vulval damage.
- similar foetal/placental appearance to bacterial abortion:
  - may have mucoid exudate and yellow leathery thickened placenta.
- vaginal discharge is variable.
- fungal hyphae identified on:
  - smear or histopathology of the placenta and foetal organs (liver).
  - may be cultured from the uterus in some cases.
- *Aspergillus* spp. is the most common fungi identified.

## Non-infectious causes of abortion

### Twinning

- historically was most common cause of abortion (now accounts for 6% of cases):
  - due to advent of ultrasound monitoring of early pregnancy.
  - See page 32.

### Umbilical cord problems

- torsion of the umbilical cord, or strangulation of the foetus by the cord (Fig. 1.59):
  - common cause of abortion (38%).
  - causes foetal asphyxia, death and necrosis.
  - increased risk of torsion where excessive cord length (breed-specific length published).
  - aborted foetus has usually undergone a degree of autolysis.

### Body pregnancy

- rare and possibly associated with placental/endometrial contact surface area and shrinkage of the placental horns, resulting in growth retardation.

### Villous atrophy

- endometrial scarring/fibrosis or lymphatic cyst formation:
  - reduce ability of placenta to closely associate with the endometrium.
  - reducing the nutrition to the foal.
- commonly seen in older mares with history of endometritis.
  - uterine biopsy is advisable to identify degree of endometrial disease and treatment.

**FIG. 1.59** Post-mortem examination of a foal aborted due to torsion of the umbilical cord.

### Foetal abnormalities

- genetic or developmental defects that are incompatible with life:
  - result in early pregnancy failure/absorption but can be aborted after 6 months gestation.

### Maternal disease

- any condition, disease or environmental circumstance that causes stress to the mare:
  - injury   - pyrexia   - endotoxaemia
  - malnutrition   - toxicosis
  - medication   - transport etc.
- placental infection/separation.
- comprehensive examination of mare and discussion with mare owner *prior* to breeding is recommended to identify any pre-existing conditions that may predispose to pregnancy failure.

## MANAGEMENT OF THE FOALING CASE

- wide variation in length of gestation (range 320–345 days).
- foetus *must* be fully mature before delivery:
  - number of physiological changes must occur prior to birth.
- 'readiness for birth' relies upon the foetus signalling it is mature:
  - associated in part with an increase in foetal cortisol levels and foetal activity.
  - e.g. final maturation of the respiratory tract occurs days prior to parturition.
- three stages of parturition:
  - 1 – build-up of coordinated uterine contractions which can take place over a few hours–days prior to delivery.
  - 2 – delivery of the foal characterised by uterine and abdominal contractions.
  - 3 – delivery of the placenta.

### Changes in the mare immediately prior to foaling.

### Endocrinology
- high levels of foetoplacental-derived oestrogen until immediately after foaling.
- progesterone levels rise during the last 20 days to a peak 5 days prior to foaling:
  - Pregnanes, although high, are decreasing during this time.
  - both drop suddenly to baseline following parturition.
- oestrogen:progestagens ratio changes stimulate uterine muscular ability prior to parturition.
- placental relaxin increases late in pregnancy leading to pelvic ligament and cervical relaxation.
- prostaglandin is produced by the foetoplacental unit:
  - slowly increases during the last trimester.
  - more rapid increase in the last few weeks leading to:
    - ◆ cervical relaxation and development of coordinated uterine contraction.

- Oxytocin is released as the foetus enters the pelvis/birth canal:
  - stimulates uterine contraction and second-stage parturition.
- after delivery, oxytocin levels drop but remain pulsatile to expel the placenta during the third stage.
- Prolactin, from the anterior pituitary, increases during late pregnancy leading to the onset of lactation.

### Physical signs
- generalised relaxation over 2 weeks in preparation for foaling:
  - tail head – tail becomes lax.
  - pelvic ligaments relax.
  - vulva softens and lengthens.
- mammary gland development begins 4–6 weeks before parturition:
  - most obvious in 2 weeks prior to parturition, especially in last 24–48 hours.
  - mare and age variable.
  - premature lactation and leakage of milk may be a sign of high-risk pregnancy.
- mammary gland secretions change from yellow, serous fluid to thick, pale-yellow colostrum:
  - loss of colostrum before birth should be noted.
  - increases the risk of neonatal disease.
- 'Wax' may be seen on the teats 24–48 hours prior to foaling but is mare variable.
- electrolyte changes within the milk can be used to determine the readiness for birth:
  - calcium – levels increase 24–48 hours before foaling (>10 mmol/l; 40 mg/dl).
  - sodium – levels drop 24–48 hours before foaling (<30 mmol/l; 30 mEq/l)
  - potassium – levels increase 24–48 hours before foaling (>35 mmol/l; 35 mEq/l)
  - sodium and potassium levels cross over 24–48 hours prior to foaling.
  - analysis is available from commercial laboratories or as a stable-side kit.

# Normal parturition

## First stage (Fig. 1.60)

- can be confused with colic as signs are similar:
  - lying down, rolling, pawing, reduced appetite and sweating.
  - frequent and small volumes of urine and faeces passed.
  - 'Flehmen' reaction.
- mild uterine contractions begin together with cervical relaxation.
- foal changes to a dorsosacral position ('diving') via:
  - own movements, uterine contractions and mare movements (rolling).
  - later the foal extends its neck and forelimbs.
- mare has control over this stage and can influence it if disturbed.
- variable time length:
  - usually 30 mins–6 hours but can be up to a few days in some cases.
- rupture of the chorioallantois (waters breaking) marks the end of stage 1.

## Second stage (Figs. 1.61–1.64)

- very rapid and active event.
- following the chorioallantois rupture:
  - amnion (white sack) and foal's front feet appear at the vulva within 5–10 mins.
    - amnion ruptures releasing yellow amniotic fluid.
  - foal is delivered within 20–30 mins (10–60 mins).
  - foal takes an active part in the delivery and foal abnormality will affect the process.
- strong abdominal contractions move the foal thorough the pelvic canal.
- mare often in lateral recumbency but maidens can foal standing up:
  - mares may get up and down repeatedly during this stage.
  - may assist in repositioning the foal and facilitating passage.
- contractions will continue until foal's hips are passed then mare generally relaxes with the hindlimbs still in the mare.
- umbilical cord continues to pulse:
  - breakage 3–4 cm from the abdominal wall occurs naturally when the foal/mare moves.
  - recoil of the blood vessels helps stop bleeding.
  - cutting with scissors should be avoided if possible.
- mare and foal should be left to bond.

## Third stage

- passage of the placenta is aided by uterine contractions and should occur within 4 hours:
  - can be associated with abdominal pain – must be differentiated from abnormalities.
- if the placenta is hanging it should be tied up above the hocks.
- once expelled, placenta must be examined for completeness and abnormalities (Fig. 1.65)

# Induction of parturition

- foetal maturity and viability must always be assessed prior to induction:
  - risk of serious complication for foal and mare including:
    - prematurity/dysmaturity
    - perinatal asphyxia syndrome.
    - retained placenta    dystocia.
- considered as a last resort option in cases such as:
  - high-risk pregnancy.
  - history of difficult foaling or abnormal foals.
  - previous injury to mare e.g. prepubic tendon rupture or fractured pelvis.
- **mare should never be induced for convenience of humans.**

## Decision-making process for induction to reduce risk of complications

- greater than 330 days' gestation.
- evidence of maternal readiness for foaling – relaxed perineum and vulva.
- foetus in normal presentation, position and posture.
- mammary secretions are consistent with foetal readiness for birth:
  - thick, yellow colostrum (can use BRIX% refractometer to assess).
- milk electrolytes should show readiness for birth:
  - calcium >10 mmol/l

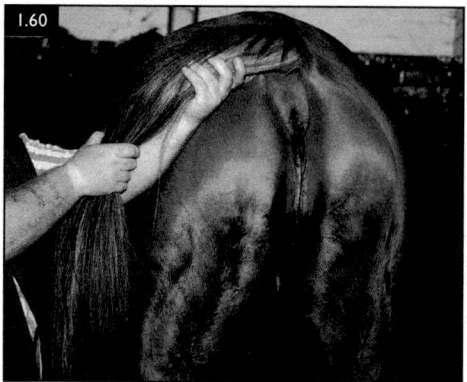

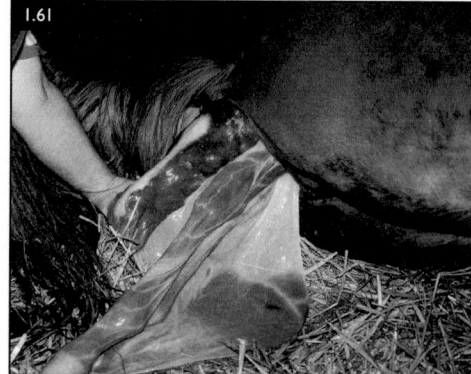

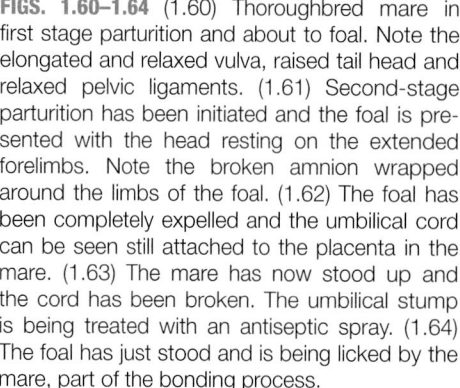

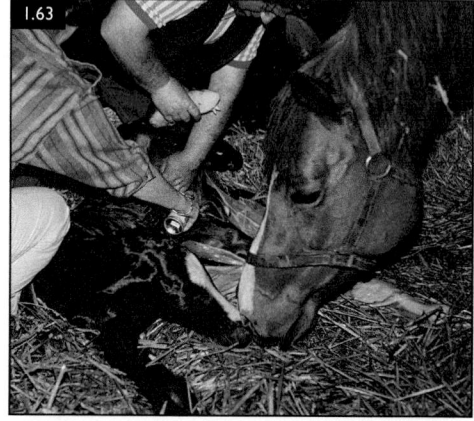

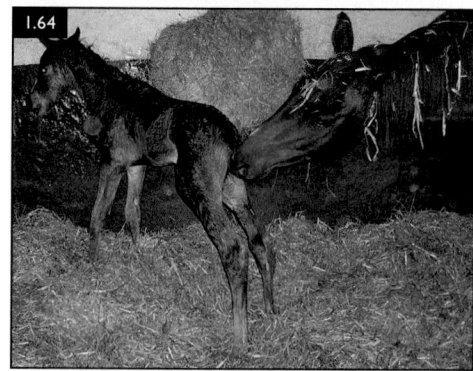

**FIGS. 1.60–1.64** (1.60) Thoroughbred mare in first stage parturition and about to foal. Note the elongated and relaxed vulva, raised tail head and relaxed pelvic ligaments. (1.61) Second-stage parturition has been initiated and the foal is presented with the head resting on the extended forelimbs. Note the broken amnion wrapped around the limbs of the foal. (1.62) The foal has been completely expelled and the umbilical cord can be seen still attached to the placenta in the mare. (1.63) The mare has now stood up and the cord has been broken. The umbilical stump is being treated with an antiseptic spray. (1.64) The foal has just stood and is being licked by the mare, part of the bonding process.

- ○ sodium <30 mmol/l
- ○ potassium >35 mmol/l.

## Methods of induction

- several methods are described each with advantages and disadvantages.

- all cases require veterinary supervision at foaling due to risk of dystocia and foetal compromise.

## Oxytocin

- method of choice in mares over 300 days gestation.

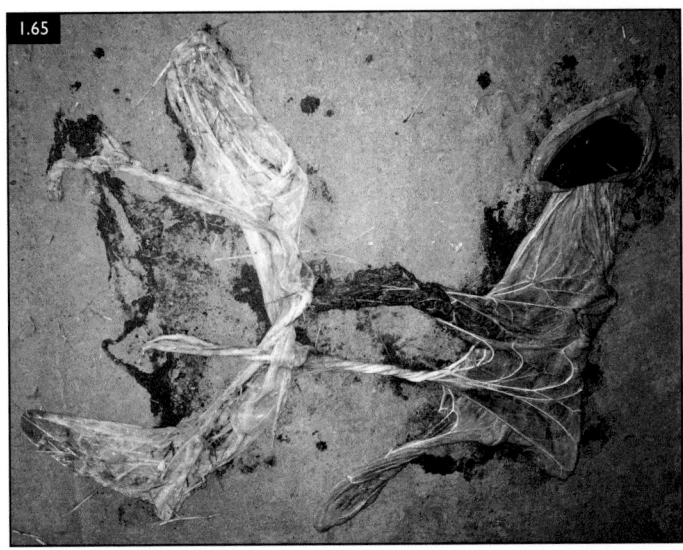

**FIG 1.65** This is a normal expulsed placenta showing the allantochorion to the right and the amnion remnants to the left.

- low dose method:
  - 2.5–10 IU i/v will induce mare to foal within 15–20 minutes.
  - can be repeated every 20–30 mins depending on progress.
- high dose method:
  - up to 20 IU i/v or via a drip (60 IU/500 ml saline) over 15 minutes.
  - foaling usually starts within 30 minutes of administration.
  - can be associated with strong vigorous abdominal contractions.
- manual cervical dilation with/without PGE/misoprostol onto the cervix should be performed to relax and dilate the cervix and reduce cervical trauma.

## Corticosteroids

- high doses of dexamethasone i/m q24h for > 4 days.
- **not generally recommended due to potential side-effects:**
  - mare does not foal at a predicable time after administration.
- used to help mature the foetus in high-risk pregnancy that might not reach normal gestational length.

## Prostaglandin $F_2\alpha$

- various forms are used but not generally effective if the mare is not ready to foal.

## DYSTOCIA

### Definition/overview

- Dystocia, or 'obstructed parturition', has a variable incidence across the equine breeds:
  - Thoroughbreds 4–5%
  - Draught and Shetlands 8–12%.
  - may be more common in the young, primiparous mare.
- most common causes are:
  - long foetal limbs and neck preventing/restricting passage through the reproductive tract.
  - foetal oversize is a rare complication in the horse compared with farm animal species.

- 'true emergency':
  - window for a successful intervention very small.
  - rapid compromise of the foetus and potential severe injury in the mare due to placental detachment and strong abdominal contractions in the mare.
- generally, if there is no progress within 15 minutes of chorioallantoic rupture:
  - veterinary examination and/or intervention is indicated.
  - slow progress of stage 2 is common in the maiden foaling mare.

## General advice to the owner of the mare with a dystocia includes

- keep the mare under control using a head collar and lead rope.
- keep the mare walking to reduce straining and impaction of the foal in the pelvic canal.
- apply a clean tail bandage.
- have a bucket of warm water and help available.

## Specific presentations that require special instruction include

- **Premature placental separation** ('red bag' delivery) (Fig 1.66):
  - placenta separates during stage 2 and does not rupture:
    - ◆ stops the oxygen supply to the foal.
  - placenta should be ruptured:
    - ◆ foal delivered as quickly as possible (take care not to damage the mare).
  - foal is at increased risk of perinatal asphyxia syndrome.
- **Upside down presentation:**
  - allow mare to get up and down as this may assist in repositioning of the foal.

**FIG 1.66** Premature placental separation or 'red bag' delivery. The chorioallantoic membrane should be broken promptly and the foal delivered.

- **Hip or shoulder 'lock':**
  - foal 'stuck' at the shoulders or hip.
  - foal should be supported and rotated slightly while pulling (hip lock) or one limb advanced at a time (shoulder lock).
- **Posterior presentation:**
  - rapid delivery of the foal is necessary as the umbilical cord will break when the foal is still in utero risking asphyxia.

## Preparation of the mare for examination

### Restraint

- foaling box should be large (4 m × 4 m), clean, dry and with a straw bed.
- experienced handler is essential.
- nose/neck twitch can help with initial restraint.
- sedation is often needed to allow examination and correction:
  - low doses of detomidine with/without butorphanol is recommended.
  - romifidine/xylazine with butorphanol are alternatives.
- short general anaesthetic helpful if the correction is prolonged or abdominal straining is problematic.

### Preparation

- tail bandage and ideally held to one side by an assistant.
- clean perineum before internal examination ideally with warm dilute antiseptic solution.
- initial internal examination with clean rectal sleeve and fresh clean obstetric lubricant keeps hands clean/dry to draw up any drugs that need to be given after the initial assessment.
- generally, mares tolerate examination well but initially stand to one side to gauge mare's response.

### Medical treatment

- sedation (see above).
- general anaesthesia (see above).
- **Clenbuterol:**
  - $\beta_2$ agonist; dose 12–15 ml i/v for 500 kg horse.

- helps reduces smooth muscle contraction/straining.
- can be enough to allow correction of the dystocia.
- not licensed for this use in some countries – potential cardiac side-effects.
- **Oxytocin:**
  - 10 IU i/m or i/v for 500 kg horse q8h.
  - used in post-partum mare to encourage placental detachment/passage and uterine involution.
  - peri-parturient mare is more receptive to oxytocin (up-regulation of oxytocin receptors around foaling).
  - complications include colic due to myometrial contractions.

## Other techniques

- **General anaesthesia:**
  - helpful in repositioning foal where abdominal straining is excessive.
  - anaesthetised mare has hindquarters elevated:
    - slope in a field
    - stable hoist
    - front loader machine.
  - takes considerable time and if a caesarean is likely to be necessary early referral is preferable.
- **Epidural anaesthesia:**
  - provides perineal and caudal reproductive tract analgesia but does *not* affect abdominal straining so is not commonly used in the foaling mare.
  - can be useful in post-foaling reconstructive surgery.

## Veterinary examination

- important that the clinician multitasks during the first few minutes when attending a dystocia:
  - brief, concise history.
  - gain control of the mare's behaviour.
  - make an initial examination:

## History

- length of second-stage parturition.
- any previous problems during parturition or pregnancy.
- length of gestation.
- normal pre-foaling signs:

- mammary gland development, waxing up, relaxed tail head etc.
- mare scanned with a single pregnancy.
- other medical problems of the mare.

## Examination

- brief assessment of mucous membranes, capillary refill, and heart rate if possible.
- internal obstetric examination (gloved and lube) to assess:
  - foal presentation, position and posture.
  - can head, feet, and limbs be palpated.
  - state of mare's reproductive tract and room for delivery or manipulation.
- foal alive:
  - does it move, finger into the mouth for chewing/suckle response.
- formulate a plan to rectify any problem.
- normal foaling:
  - **presentation** – anterior longitudinal.
  - **position** – dorsosacral.
  - **posture** – extended head resting on the carpi of the extended forelimbs.
- any deviation from the above indicates a problem that must be corrected.

## Foaling technique

- time is limited and compromise to foal and mare may result in death of one or both.
- after initial assessment a rapid decision must be made with the owner as to how to proceed.
- 4 broad courses of action are available:

1. **deliver the foal after a quick and simple manipulation.**
2. **prolonged manipulation is required to deliver a live or dead foal:**
   - early decision made to treat on-farm or refer.
   - on-farm:
     - sedation/GA, correction is made, and the foal delivered.
     - complications include foal death, foal compromise, and mare reproductive tract trauma.
     - referral after attempted correction is possible:
       - likely to result in death of foal and trauma to mare's tract.

3. **referral for caesarean section:**
   ○ diagnosis of severe malposition/malposture/deformity.
   ○ transportation must be done rapidly but safely for foal and mare welfare.
   ○ foal can be intubated *in utero* via a cuffed nasotracheal tube and oxygen provided during travelling as placental function is likely to be compromised.
4. **foal already dead:**
   ○ lubrication pumped into the uterus, if the reproductive tract is dry, and the foal delivered.
   ○ foetotomy (usually partial 2–3 cuts) can be performed and the sections delivered:
     ◆ care must be taken to protect the reproductive tract, especially the cervix.
     ◆ use the correct and appropriate equipment.
     ◆ referral may be appropriate.

## Manipulation technique

- apply foaling ropes to limbs (above the fetlock joint) and head (behind ears and above mandible) as soon as possible:
  ○ 3 different coloured ropes to help identify which part of the foal attached to.
- plenty of lubricant and use stomach pump and tube to infuse the uterus:
  ○ can be watered down to improve the slip.
- when repelling the foal back into the mare do so with firm and constant pressure.
- cup the hoof in the palm of the hand to correct limb malpostures and extend the limb by moving the foot axially/medially to avoid traumatising the uterus/cervix.
- pull the foal only when the mare is pushing.
- maximum of 2–3 people should apply traction.
- pulling one leg, then the other, then the head in rotation reduces the width of the foal and can assist in 'walking' the foal out.
- consider referral and/or general anaesthesia early if prolonged manipulation is anticipated.
- referral is not an option:
  ○ mare may be anaesthetised and a terminal caesarean section performed in an attempt to save the foal.

○ mare must be euthanased immediately the foal is delivered.
○ foal is likely to require resuscitation/intensive treatment.

# Specific dystocia presentations

- **Incomplete elbow extension:**
  ○ repel the foal and extend the limbs.
- **Foot–nape** – front foot over the head:
  ○ easy to correct and deliver.
  ○ not corrected leads to high risk of rectovaginal fistula or perineal laceration.
- **Ventral head flexion** – head back:
  ○ apply ropes to front limbs, repel foal, reach and pull head up.
  ○ apply rope to head and deliver foal.
    ◆ head may be difficult to reach in large mares.
- **Carpal flexion** – one or both front limbs:
  ○ no hoof palpated – only the carpus.
  ○ repel the foal, cup the hoof, and correct posture by moving hoof medially.
  ○ correction is not possible, consider that the foal has carpal contracture.
- **Carpal flexion due to carpal contracture:**
  ○ as above but can be difficult to correct:
  ○ cervical damage is likely so consider caesarean section if alive or partial foetotomy if foal is dead.
- **'Dog sitter':**
  ○ difficult to identify and correct.
    ◆ identify by lack of progress even though head/front limbs correctly presented.
  ○ hindlimbs get caught under the pelvic brim.
  ○ consider referral or GA.
- **Shoulder flexion:**
  ○ may affect one or both limbs.
  ○ repel foal and correct each limb to carpal flexion initially.
  ○ then proceed as above to avoid foal entering the pelvic canal before fully corrected.
- **Posterior presentation** (backwards presentation):
  ○ rare, umbilical cord is under tension and will break with foal *in utero*.
  ○ rapid delivery is needed and monitoring foal for signs of asphyxia.

- **Hindlimb forward** ('partial dog sitter'):
  - may be deliverable with one hindlimb along body.
  - cervical damage and/or locking at the chest may occur.
- **Transverse presentation:**
  - correction to anterior presentation may be possible but frequently limbs cannot be reached.
  - mare should be referred for caesarean section.
- **Congenital abnormalities:**
  - relatively rare but often lead to dystocia if involve limb flexural issues or cervical vertebrae deformity ('wry neck') (Fig. 1.67)
  - some of these cases will require caesarean section for delivery of the foal.

## Potential complications

- **Excessively long gestation:**
  - some pregnancies run over 12 months:
  - may result in a normal foal.
  - commonly the foal is small and weak, requiring special attention.
  - often seen in older mares possibly due to uterine/placental issues causing poor foetal growth and development.
- **Fractured pelvis of mare:**
  - mares with healed pelvic fractures can have a normal breeding life.
  - require monitoring during foaling in case of a problem (altered pelvic anatomy).
  - these mares should be examined before breeding:
    - ◆ significant narrowing/alteration in pelvic dimensions should preclude breeding.
- **Rib fractures of foal:**
  - during parturition, foal's ribs can be fractured as the chest squeezes through the pelvis.
  - veterinary assessment of every foal 24–48 hours after birth is recommended.
- **Perineal melanomas:**
  - often incidental but if numerous can be painful and become traumatised at foaling.
  - where complications are likely, the mare should not be naturally bred, and the use of embryo transfer/oocyte collection considered.

**FIG. 1.67** A newborn Thoroughbred foal showing contracted foal syndrome, including bilateral carpal flexural deformities and wrynose. The foal presented as a dystocia in the mare because of a partial carpal flexion deformity.

## Caesarean section

- indicated when a foal cannot be delivered per vagina:
  - congenital deformity.
  - foetal oversize.
  - severe malpresentation/malposture.
  - where correction of dystocia may compromise future fertility due to cervical or reproductive trauma.
- prompt referral for surgery greatly increases the success of a live mare and foal:
  - ideally, the surgery should be within 1 hour of the onset of stage 2 parturition.
- experienced surgical team and postoperative care is essential.
- ventral midline laparotomy is usual although a paramedian approach has been used in mares with previous midline surgery.
- complications include:
  - postoperative uterine haemorrhage.
  - retained placenta.
  - abdominal incisional infection/ breakdown.
  - laminitis/metritis  ○ shock.
  - peritonitis  ○ adhesions.
- usually rest mare from breeding until the following season.
- foal should be considered high risk and given appropriate treatment, resuscitation and intensive care.

## THE POST-PARTUM PERIOD

### Introduction

- mare rapidly recovers following an uncomplicated foaling, returning to normal behaviour and looking after the foal within an hour of delivery.
- uterine contractions continue in the post-partum period to expel the uterine debris:
  - fluid, bacterial contamination and cellular material.
- uterine immune response resolves any bacterial contamination.
- uterine involution returns the uterus to a non-pregnancy size over the first 2 weeks.
- all these processes can be encouraged by paddock exercise, foal suckling (oxytocin release):
  - delayed in cases of dystocia, internal manipulations or where post-partum management is compromised.

### Foal heat

- return to ovarian cyclicity begins shortly after foaling.
- first oestrus period is termed 'foal heat' and occurs at around 6–14 days post-partum.
  - first ovulation post-partum occurs at 5–15 days.
  - these timings decrease in the summer months.
- first oestrus period can be 'silent' or unobserved due to overriding maternal behaviour.
- mares should be examined at 9–10 days post foaling to assess the recovery process:
  - bacterial swab can be taken to ensure successful cleansing of uterus.
  - any vulval damage can be assessed and repaired as necessary.

- viewed as a uterine cleaning cycle and breeding at this time only considered in cases of:
  - normal foaling process
  - normal post-partum management.
  - adequate uterine involution
  - clean uterine smear and culture.
- breeding is generally delayed until the second oestrus post-partum (28–30 days).
- fertility at foal heat is considered good if breeding is delayed until after day 10:
  - slight increase in early pregnancy loss following foal-heat breeding.
- option to 'short cycle' the mare using prostaglandin injection at 19-20 days (or 5-7 days after foal-heat ovulation) is a method of getting the mare to foal earlier in the subsequent year while giving her time to recover after foaling.

### Post partum complications

- complications should be suspected and investigated if the following occurs:
  - abdominal pain continues or worsens.
  - reduced appetite.
  - sweating    o pyrexia.
  - increased heart/respiratory rates.
  - pale/injected mucous membranes.
  - no interest in the foal
- post-partum pain related to uterine contractions:
  - most common in primiparous mares.
    - moderate increased pulse rate with mild sweating.
    - subsides in 1–2 hours.
    - differentiate from other more serious problems.
- maternal illness can reduce milk production:
  - compromise foal nutrition, hydration and growth.

## CONDITIONS ASSOCIATED WITH FOALING (POST-PARTUM)

### Uterine artery rupture

#### Definition/overview

- well recognised but uncommon condition of post-foaling mare.

- rarely occurs in late-pregnant mare.
- rupture/bleeding of uterine artery into mesometrium (broad ligament) or peritoneal cavity.

- clinical signs vary from mild/moderate pain to shock and death.
- symptomatic treatment.

## Aetiology/pathophysiology

- Middle uterine, utero-ovarian or external iliac arteries are the most affected.
- unknown aetiology.
- more common in older, multiparous mares:
  - possibly due to previous stretching and progressive weakening of arteries and supportive structures.
- possible predilection to right-side rupture.
- haemorrhage is initially contained within the broad ligament:
  - if remains within this structure:
    - haematoma formed which helps stop the bleeding.
  - haemorrhage continues or broad ligament ruptures (Fig. 1.68):
    - bleeding into peritoneal cavity.
    - systemic shock.
    - severe cases exsanguination and death occur rapidly or over a few hours.

## Clinical presentation

- mild to severe abdominal pain depending on degree of haemorrhage.
- Flehmen response, sweating, anxiety, weakness and pale mucous membranes.
- perineal swelling may occur from caudal extension of haemorrhage.
- severe or unrestricted bleeding into peritoneal cavity signs include:
  - above plus rapid heart rate, thready pulse, prolonged capillary refill time.
  - collapse/recumbency, unable to stand following foaling and sudden death.

## Diagnosis

- clinical signs and timing of presentation are highly suggestive.
- careful rectal palpation in mild cases for presence of broad ligament haematoma:
  - size variable from apple to melon.
  - transrectal ultrasound confirms free blood within the structure (Fig. 1.69).
- transabdominal ultrasound will show free blood in the cavity (extra-ligament bleeding).

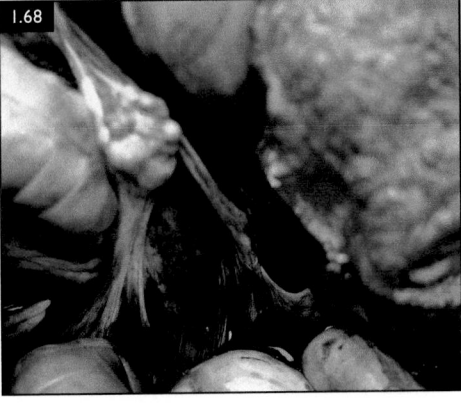

**FIG. 1.68** Post-mortem view of the ovary and broad ligament of a mare that died because of a uterine artery rupture that bled from the ligament into the peritoneum.

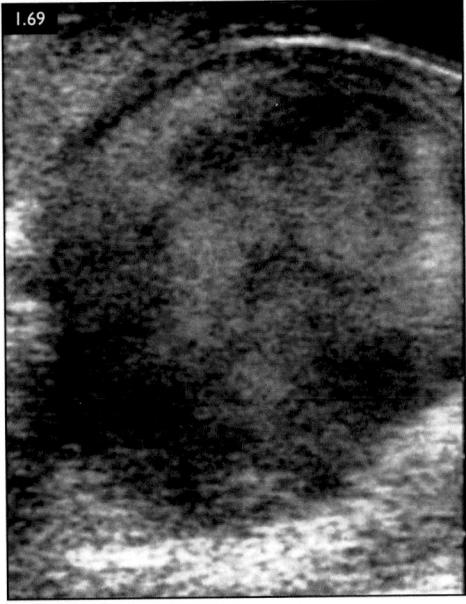

**FIG. 1.69** Transrectal ultrasonographic view of a haematoma within the broad ligament 24 hours post-partum.

- blood sample analysis of variable use in early stages (less than 12 hrs.).

## Management

- keep mare quiet and minimise stress.
- strict box rest:
  - clot formation takes approximately 7 days.
  - 14-day minimum box rest period.

- keep mare and foal together to minimise stress:
  - beware of mare collapsing/injuring foal.
- analgesics such as flunixin meglumine.
- avoid sedatives.
- anti-fibrinolytic drugs (tranexamic acid or aminocaproic acid):
  - encourage clot formation/stabilisation but effectiveness unknown.
- corticosteroids, broad-spectrum antibiotics, antioxidant medications are all recommended.
- intravenous fluids can be used for circulatory shock:
  - 2–4 litres of warm hypertonic saline (7%) or hetastarch.
  - followed by polyionic crystalloid solution to restore circulatory function.
  - blood transfusion and plasma have also been used.
  - **fluid therapy may cause disturbance of the clot and reactivation of bleeding.**
- surgical ligation of the vessels is theoretically possible but very difficult to perform.
- foal nutrition should be supported.

## Prognosis

- bleeding is contained within the broad ligament – the mare should survive.
- intra-abdominal bleeding carries a poor to hopeless prognosis.
- mares examined at the 30-day heat period with assessment of the haematoma:
  - breeding can occur if organised.
  - larger clots may delay breeding until the following year.

# Uterine rupture

## Definition/overview

- uncommon but usually secondary to dystocia:
  - can occur in normal foaling – damage from hindlimbs of foal during parturition.
- partial tears may not be detected and heal spontaneously as uterus involutes.
- full thickness tears are associated with pain and peritoneal haemorrhage and contamination.
- death can occur due to peritonitis or intestinal complications.

- surgical repair and peritoneal lavage are essential for treating full thickness tears.

## Aetiology/pathophysiology

- usually, secondary to other conditions:
  - hydrops
  - uterine torsion
  - dystocia management.
  - generally in ventral body.
  - foetal movements (hindlimb) can cause tears, usually at tip of the pregnant horn.
- full thickness tears lead to abdominal contamination.
- abdominal visceral herniation into the uterus or peritoneal haemorrhage also possible.
- occur at parturition and also rarely in late pregnancy:
  - foetus may move into the peritoneal cavity in the latter case.

## Clinical presentation

- partial tears are often not detected.
- abdominal haemorrhage may present with shock and rapid death.
- extra-uterine late pregnancies may present with abdominal pain and intestinal complications.
- full thickness tears may heal spontaneously or present with peritonitis:
  - abdominal pain and guarding
  - pyrexia
  - depression
  - signs of endotoxaemia/septic shock.

## Diagnosis

- challenging to confirm suspicion.
- manual examination of the uterus per vagina, by rectal palpation and by transrectal ultrasonography can identify the site of injury in many cases:
  - carry out carefully to minimise further damage.
  - should be carried out routinely in all dystocia cases.
- post-partum lavage fluid may not be fully retrieved suggesting leakage out of uterus.
- transabdominal ultrasonography and peritoneal fluid analysis to identify peritoneal contamination.
- laparoscopy.

## Management

- often presented late when secondary complications have presented e.g. peritonitis.
- partial tears often require no treatment.
- **uterine lavage should not be performed.**
- cases presenting with peritonitis:
  - surgical repair – midline laparotomy or flank laparoscopic approach (Fig. 1.70).
  - peritoneal lavage.
  - intravenous fluid therapy and prophylaxis treatment for laminitis.
- all mares with partial or complete tears should be treated with:
  - oxytocin (10–20 IU i/m q2–4h) to encourage involution.
  - antiendotoxic NSAIDs.
  - other anti-endotoxic drugs.
  - broad-spectrum antibiotics.
- foal nutrition should be supported.

## Prognosis

- guarded to grave depending upon extent of damage and secondary complications:
  - adhesions and uterine defects can remain.
- partial thickness tears can be bred following resolution, typically after 60 days.
- full thickness tears should be rested until the following year.

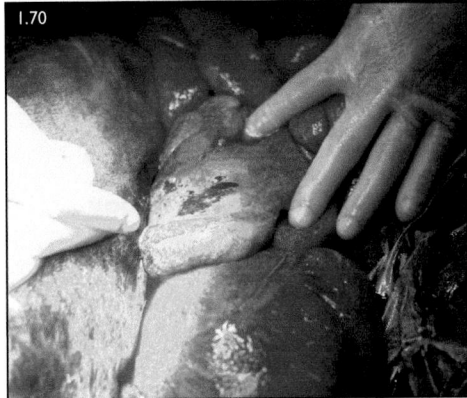

**FIG. 1.70** View at laparotomy showing a small tear in the uterus of a recently parturient mare. Note the inflamed serosa of the uterus subsequent to generalised septic peritonitis.

# Uterine prolapse

## Definition/overview

- uncommon condition in the post parturient mare where there has been:
  - dystocia.
  - obstetrical manipulation/extraction.
  - excessive straining post-partum because of treatment of retained foetal membranes.
- complete prolapse life-threatening emergency.
- partial prolapse or inversion of the tip of uterine horn can be noted at a post-partum exam.
- treatment – replacement under sedation and epidural anaesthesia.

## Aetiology/pathophysiology

- following:
  - dystocia, especially if prolonged/ forceful manipulation required.
  - excessively rapid foaling.
  - excessively rapid/forceful removal of placenta.
  - late-gestation abortion.
  - any condition causing excessive abdominal straining post-partum:
    - retained placenta
    - vulval/vaginal trauma.
- uterus may have some damage following prolapse.
- uterine blood vessels can be damaged/ ruptured leading to exsanguination and death.

## Clinical presentation

- everted uterus seen at vulval lips (Fig. 1.71) from hours to days following foaling.
  - bright/dark red due to haemorrhage or blood stasis.
  - varying degrees of damage, friability and infection which negatively effects fertility.
  - placenta may still be attached.
- rapid development of shock and endotoxaemia may occur.
- uterine artery rupture may present with colic or sudden death.
- can be accompanied by bladder eversion, uterine rupture and intestinal herniation.
- invagination of the uterine horn can be subclinical or present as abdominal pain.

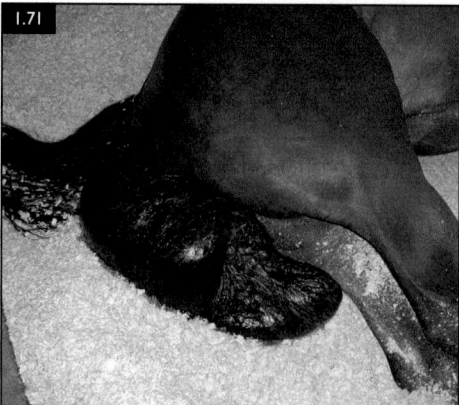

**FIG. 1.71** A uterine prolapse post-partum in a mare that was subsequently euthanased.

## Diagnosis

- clinical presentation.
- rectal/vaginal palpation :
  - invagination of the horn detected on rectal palpation as a thickened, oedematous and shortened uterine horn.

## Management

- complete prolapse is emergency.
- protect the uterus from further trauma and damage.
- restrain and keep the mare quiet to prevent further uterine damage:
  - sedatives/analgesics are often necessary but must be used with care.
  - **can lead to mare collapse due to induced hypovolaemia.**
- general anaesthesia to allow the hindquarters to be raised may be necessary.
- broad-spectrum antibiotics (including metronidazole) and NSAIDs (flunixin meglumine) prior to replacement to treat endotoxaemia/metritis.
- replacement of the uterus:
  - place on clean plastic sheet/tray and elevate to vulva to improve circulation.
  - clean uterus with warm, isotonic saline.
  - epidural anaesthesia – onset of action is 20 mins.
  - identify and treat damaged areas (absorbable sutures).
  - remove any placenta that is present and that can be easily detached.

- carefully replace – starting with the vagina.
  - ◆ use flat hands/fists and avoid finger pressure as friable uterus may tear.
- when back in position ensure tips are not invaginated:
  - ◆ manual palpation or if out of reach:
    - – fill uterus with warm isotonic fluid and confirm by rectal palpation.
    - – siphoning of the fluid will reduce contamination.
- low dose oxytocin (10 IU i/m q2–4h) will encourage uterine involution.
- retained placenta requires specific treatment.
- intrauterine antibiotic infusions can be used.
- some clinicians advocate suturing the vulva to prevent pneumovagina and further straining.
- recurrence unlikely if the uterus is completely reduced, including the horns.

## Prognosis

- varies from guarded to grave depending on degree of damage and secondary problems.
- subsequent metritis is common:
  - requires intensive treatment and may impact future fertility.
- no increased incidence in subsequent pregnancies.

# Bladder eversion and prolapse

## Definition/overview

- rare but more common in larger breeds, particularly draughts.

## Aetiology/pathophysiology

- usually occurs post-partum due to straining:
  - reproductive tract trauma, dystocia, retained placenta, intestinal impaction etc.
  - occasionally seen in prepartum mares
- Prolapse – bladder forced through tear in ventral vaginal wall.
- Eversion – bladder everted through urethral opening.

## Clinical presentation

- Prolapse – serosal surface of bladder usually exposed at vulval lips.
- Eversion – mucosal surface is presented with ureteral opening visible (Fig. 1.72).

## Management

- treatment is usually undertaken under epidural anaesthesia and standing sedation.
- **Prolapse:**
  - empty and clean the bladder.
  - replace into abdomen through the vaginal wall tear.
  - repair vaginal tear.
  - long-handled instruments and speculum for visualisation are essential.
- **Eversion:**
  - empty and clean bladder.
  - any defects repaired, bladder lubricated and replaced through urethral opening.
  - shrink an oedematous bladder by wrapping in cool, dextrose-soaked towels.
  - difficult to replace if small intestine lies behind the bladder or urethral opening is narrow.

- following replacement, straining must be limited to prevent recurrence:
  - ensure fully replaced.
  - epidural anaesthesia.
  - NSAIDs and broad-spectrum antibiotics.
- in-dwelling urinary catheter should be placed:
  - if causes straining should be removed.

## Prognosis

- depends on:
  - degree of damage.
  - ease of replacement.
  - primary cause of straining.
  - whether surgical repair to the urethra is required.
- recurrence is possible.
- increased incidence of urinary incontinence.

## Rectal prolapse

### Definition/overview (Fig. 1.73)

(see also **Book 3**)

- associated with conditions that cause increased abdominal pressure/straining:
  - dystocia.
  - retained placenta.
  - reproductive tract trauma.

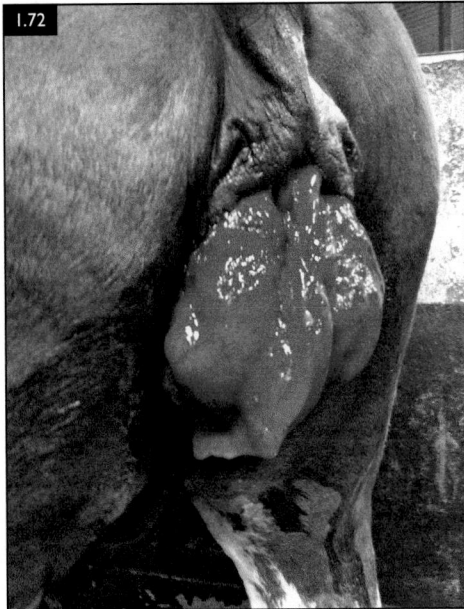

**FIG. 1.72** A mare with bladder eversion post foaling.

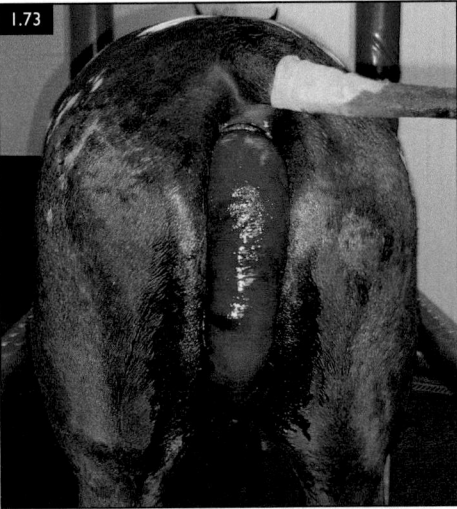

**FIG. 1.73** A type 4 rectal prolapse in a recently foaled broodmare. The prolapse contained small intestine and damaged mesentery and was not amenable to repair.

# Premature placental separation

## Definition/overview

- uncommon but emergency presentation at foaling – 'Red bag'.
- allantochorion separates from the endometrium before the foal breaks the membrane and enters the pelvic canal.
- foal is contained within the placenta and mare attempts to deliver them as a single unit.
- immediate intervention needed to break the allantochorion and deliver the foal.
- foal at risk of asphyxia and requires intensive treatment.

## Aetiology/pathophysiology

- often secondary to other problems, particularly the placenta:
  - ascending infectious placentitis.
- increased risk of occurrence with:
  - induction of parturition
  - late gestational stress/colic
  - fescue toxicosis.

## Clinical presentation

- placental infection may present with a late gestational vaginal discharge.
- occasionally vaginal bleeding for a few days prior to parturition.
- at parturition:
  - **red, thick, velvety membrane (allantochorion) at vulva at start of second stage.**
  - not usual amnion (whitish, thin membrane).
- no release of the placental fluid ('breaking of the waters').

## Management

- blood supply to the foal stops when the placenta separates.
- allantochorion must be ruptured quickly:
  - manually (normal thickness) or using scissors (if thickened e.g. placentitis).
  - ideally at the cervical star.
- deliver foal as rapidly as possible.
- resuscitation of foal – oxygen per nasum used if available.
- foals are considered 'high risk':
  - premature, dysmature, or maladjusted and/or septic.
  - treated and monitored for complications.
- examine placenta and submit samples for histopathology if abnormal.

## Prognosis

- variable depending on cause, degree of placental compromise and speed of correction and delivery.
- neonatal intensive care can be extensive and costly.
- mare with previous placentitis, should be monitored in late pregnancy by placental assessment etc., and with special attention at future foalings.

# Retained foetal membranes (placenta)

## Definition/overview

- common post-partum complication.
- normally, the placenta is passed 1–3 hours after birth (third stage parturition):
  - considerable individual variation without clinical significance.
  - multifactorial aetiology.
- serious complications can occur, particularly in draught/heavy horse breeds.
- treatment is aimed at:
  - encourage release of the placenta.
  - prevention/treatment of secondary complications.
- prognosis is fair–good in most cases if prompt and effective treatment.

## Aetiology/pathophysiology

- any factor/condition that affects uterine motility/uterine maturation pre-foaling:
  - dystocia, especially after extensive manipulations.
  - induction of parturition.
  - premature delivery and abortion.
  - caesarean section.
  - electrolyte imbalances (selenium, calcium/phosphorous).
  - uterine inertia/fatigue:
    - post twin or hydrops pregnancy.
    - inadequate exercise and obesity.
    - general anaesthetic.
  - placentitis.

- incidence of 2–10% in the general population but higher in:
  - draught breeds
  - older multiparous mares.
  - mares that have retained a placenta previously.
- non-pregnant uterine horn/tip is most frequently involved.
- mechanism that interferes with normal separation is unknown.
- complications, particularly in larger, heavy mares are associated with:
  - bacterial proliferation and autolytic enzyme build-up in the uterus.
  - subsequent systemic absorption (bacteraemia, septicaemia, endotoxaemia).
  - severe acute laminitis.
  - uterine prolapse.

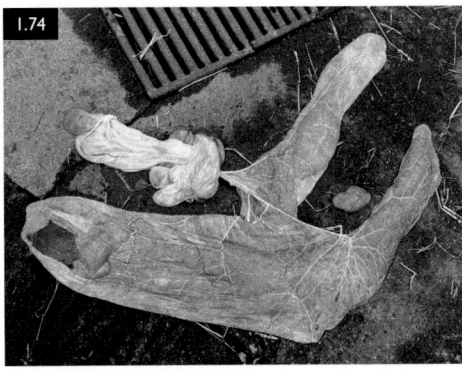

**FIG 1.74** Placenta laid out following delivery. The thin non-pregnant horn was retained and removed shortly after the main placenta was passed.

## Clinical presentation

- placenta not passed after 5–6 hours following foaling:
  - note breed variation:
    - at-risk breeds, placenta passed within 3 hours maximum post foaling.
- placenta usually hanging from the vulva or fully retained in the uterus.
- abdominal pain (variable/mild) may be present due to uterine contractions (third stage):
  - often pass the membranes quickly.
  - true retention cases have no pain.

## Diagnosis

- clinical signs:
  - full physical examination carried out to detect any complications:
  - with any complications:
    - samples for bacterial culture/ sensitivity from uterus and blood.
    - full haemotology and biochemistry blood analysis.
- placenta should be examined following passage by laying in a 'F' shape and account for all parts (Fig 1.74):
  - occasionally, the hippomane may be retained.
    - consider if mare presents for metritis but placenta confirmed as passed completely.
    - note the hippomane is not always detected post foaling.

## Management

- approach to retained foetal membranes varies considerably depending on the duration of membrane retention and the presence/absence of metritis with septicaemia.
- placenta tied up to 'hock level' after stage 2 parturition to encourage passage and avoid trauma.
- accelerated release of foetal membranes best achieved in the first 6–8 hours post-partum:
  - Oxytocin:
    - 10 IU i/v /500 kg
    - 10-20 IU i/m /500 kg.
    - via i/v drip over 15 mins (various concentrations).
    - can be repeated every 20–60 minutes depending on response.
    - manual removal should be attempted if not successful.
    - mild colic signs can be caused by bolus oxytocin.
- manual manipulation methods:
  - twisting of the allantochorion on itself to encourage release.
  - distending allantochorion (up to 10 l warm saline/water) via a sterile stomach tube:
    - membrane then tied closed for up to 30 minutes to stretch the uterus.
    - +/- low dose oxytocin.
    - distension encourages release within 30 mins.

- catheterise umbilical blood vessel with a foal stomach tube and infuse water from stomach pump/tap:
  - ◆ mare may become mildly uncomfortable towards the end of the infusion.
  - ◆ placenta usually passed within 5–10 mins.
- complications with manual removal and therefore not generally advised:
  - tearing the placenta and leaving fragments in the uterus.
  - tearing at the microcotyledon level and leaving microvilli to autolyse.
  - haemorrhage.
  - uterine damage, fibrosis and infection.
  - uterine horn invagination/uterine prolapse.
- preventative measures if placenta retained after 8 hours or 3 hours in draught breeds:
  - broad-spectrum antibiotics:
    - ◆ potentiated sulphonamides or penicillin/gentamicin
    - ◆ +/- metronidazole.
  - NSAIDs (flunixin meglumine).
  - intravenous fluids +/- electrolytes.
  - uterine lavage with isotonic fluids (saline/water with added salt) until fluid clear:
    - ◆ repeated once to twice daily until resolution.
  - use of uterine antibiotics is controversial.
  - Oxytocin 10–20 IU i/m q2–4h.
  - tetanus prophylaxis.
  - frog supports and a deep bed.
  - support foal nutrition.

## Prognosis

- depends on length of retention and breed of horse:
  - fair–good prognosis generally if prompt treatment.
  - poor–grave if complications occur or chronic cases.
  - draught breeds are particularly at risk.
- future fertility can be affected if aggressive removal or chronic cases:
  - check uterine involution at foal heat.
  - confirm no uterine infection with endometrial swab/smear.

# Cervical injury

## Definition/overview

- usually occur at parturition.
- mild to severe damage possible:
  - mucosal defect.
  - adhesion.
  - partial/full thickness lacerations.
- damage may go unnoticed until failure to get in foal.
- diagnosis by careful manual palpation, ideally during dioestrus.
- treatment and prognosis depend on degree of damage.

## Aetiology/pathophysiology

- trauma occurs during parturition particularly with:
  - dystocia
  - inadequate lubrication.
  - foetotomy
  - failure of cervical relaxation (maiden mare).
  - manual manipulation
  - occasionally in normal parturition.
- mucosal injuries are the least serious and heal by epithelialisation:
  - delayed healing may lead to fibrous adhesions.
- fibrosis/scarring of the cervix can lead to adhesions/distortion/obstruction of the canal.
- adhesions within the uterus can also be encountered particularly following the intrauterine infusion of caustic substances e.g. strong iodine.
- cervical incompetence can occur when the muscular layers are damaged without mucosal involvement.
- cervical lacerations (mucosal and muscular) are usually wedge-shaped (base at the external os).
- injuries that affect cervical function can lead to sub/infertility due to:
  - canal obstruction
  - cervical incompetence.
- congenital abnormalities/hypoplasia/ malformation of the cervix and neoplasia are rare.

## Clinical presentation

- often unnoticed at parturition.

- may be identified at foal heat or during investigation of the result of the cervical dysfunction:
  - chronic endometritis.
  - failure to conceive.
  - pregnancy failure often beyond 4–5 months.

## Diagnosis

- examine the cervix via a speculum:
  - open cervix during dioestrus indicates abnormality.
- manual palpation of external os using gloved lubricated thumb and forefinger.
  - easier to assess during dioestrus (progesterone dominance) when cervix closed.
- investigation following suspected trauma at dystocia is advised (within 3–5 days).
- assess the rest of the reproductive tract as well in severe injuries.

## Management

- mucosal damage:
  - flush affected area and apply emollient cream to reduce adhesion formation.
    - most lesions epithelialise by the first post-partum oestrus, unless very large.
- recently formed adhesions can be manually broken down:
  - followed by twice daily cervical manipulation and antibiotic/ corticosteroid cream application for 7–10 days.
- chronic adhesions require surgical resection with high likelihood of recurrence.
- partial thickness or small full thickness tears can heal spontaneously by second intention:
  - cervical integrity should be assessed during second dioestrus phase post-partum.
- full thickness tears of greater than half length of cervix require:
  - 3-layer surgical repair under standing sedation no earlier than 4–6 weeks post-partum.
  - inflammation subsided and healthy granulation tissue present.
- postoperative antibiotics and daily application of antibiotic/corticosteroid cream to cervix for 3–5 days to minimise adhesion formation.
- following repair natural cover should be delayed for 90 days and AI for 30 days.
- wound breakdown is common.
- all mares with cervical injury should be monitored in subsequent pregnancies:
  - increased risk of cervical incompetence.
  - development of ascending infection/ placentitis.
  - exogenous progesterone to aid cervical closure is advised.
- cervical incompetence is difficult to treat but usually by surgical repair.
- severe cervical damage, not amenable to repair:
  - can be treated with cervical cerclage after early diagnosis of pregnancy.

## Prognosis

- depends on degree of damage and when identified.
  - adhesions and chronic cases carry a guarded prognosis.
  - mucosal/small partial tears carry fair– guarded prognosis.
- surgical repair is difficult, and complications are common.
  - fertility post repair is fair, but recurrence is common.
- development of placentitis with abortion/ placental compromise is a risk.

# Metritis–Laminitis– Septicaemia complex

## Definition/overview

- **medical emergency.**
- absorption of uterine bacterial toxins leading to septicaemia/laminitis.
- associated with retained placenta or uterine infection following foaling.
- increased incidence in draught/heavy breeds of horse (Fig. 1.75).
- progression of clinical signs can be rapid, and death can occur in severe cases.

## Aetiology/pathophysiology

- gross uterine contamination can occur during dystocia or with retained placenta:
  - leading to bacterial infection of endo/ myometrium and even serosal layers.

**FIG. 1.75** A Shire mare that has developed metritis–laminitis–septicaemia complex after a retained placenta. The mare is reluctant to stand and is on a deep shavings bed, partially supported in slings.

- large volume of purulent uterine fluid can accumulate without noticeable vaginal discharge.
- absorption of bacteria and toxins causes septicaemia with/without toxaemia.
- peripheral vascular effects of toxins may result in acute and severe laminitis.
- Gram-negative bacteria are the most common isolates (*Escherichia coli*).

## Clinical presentation

- following foaling/dystocia/retained placenta:
  - rapidly within 12 hours or a slower onset (up to 10 days).
- pyrexia • depression/dullness.
- vaginal discharge (inconsistent) – reddish/brown and foetid.
- abdominal pain
- decreased milk production.
- increased heart and respiratory rates.
- laminitis within 1–5 days - acute, digital pulses, 2–4 feet involved.
- severe septicaemia and endotoxaemia ensue with injected mucous membranes.
- circulatory collapse and death.

## Differential diagnosis

- colic • large intestine torsion.
- uterine rupture
- uterine artery haemorrhage.

## Diagnosis

- history and clinical signs.

- rectal palpation and transrectal ultrasound:
  - enlarged, doughy uterus with large volume of foetid/hyperechoic fluid.
- culture/sensitivity/cytology of uterine fluid.
- blood samples for full haematology/biochemistry – neutropaenia and toxic changes.
- palpation and radiography of the feet.

## Management

- immediate uterine lavage:
  - warmed (sterile) saline (some clinicians use 0.5% dilute povidone–iodine solution).
  - via sterile nasogastric, wide-bore tube until uterus comfortably distended.
  - fluid/debris/toxins removed – process repeated until fluid returns clear.
- uterine infusion of antibiotic solution (effective against gram-negative bacteria).
- repeat lavage and infusion 1–2 times daily until initial flush is clean with no debris.
- monitor response to treatment by rectal palpation and ultrasonography.
- broad-spectrum antibiotics (e.g. penicillin/gentamicin; oxytetracycline; +/- metronidazole).
- NSAIDs (Flunixin meglumine).
- Oxytocin 10–20 IU i/m q2–4h to aid uterine clearance.
- intravenous fluids +/- electrolytes as appropriate.

- early treatment or preventative measures for acute, severe laminitis.
- foal nutrition should be supported.

## Prognosis

- often pathology is advanced when presented, which makes the prognosis poor.
- early recognition and treatment are essential to improve chances for a successful outcome.
- treatment can be prolonged and expensive.
- severe laminitis carries a poor prognosis.

## Perineal lacerations

### Definition/overview

- common injuries, especially in the primiparous mare.
- often associated with foetal malpresentation but can occur during normal foaling.
- first-, second- and third-degree lacerations are recognised.
- visual and manual examination required to assess injury.
- treatment varies with the degree of damage.

### Aetiology/pathophysiology

- primiparous mares at increased risk due to:
  - 'tighter' reproductive tract and increased risk of foetal oversize.

- presence of hymen remnant/ vestibulovaginal sphincter:
  - increased risk of foal's foot getting caught during foaling causing damage to dorsal vaginal wall/ rectum.
- certain malpresentations e.g. foot–nape associated with damage.
- degree of damage is variable depending on whether the foal's feet are withdrawn/ corrected after the damage occurs or they continue on the same path.
- damage can also occur to the lateral vaginal wall, caudal rectum and anal sphincter.

### Clinical presentation

- **First-degree** laceration:
  - involve mucosa of vestibule and dorsal vulval commissure (Fig 1.76a).
- **Second**-degree laceration:
  - affects the mucosa, submucosa of vestibule, the dorsal vulva and some perineal muscle (Fig. 1.76b).
- **Third**-degree laceration:
  - involves all layers of the dorsal vestibule, perineal body, caudal rectum and external anus (Fig. 1.76c).
- all this damage is retroperitoneal:
  - rare cases, may also involve more cranial tissues:
    - possibility of peritoneal contamination and peritonitis.
- **Rectovaginal fistula:**
  - formed when foal's feet are withdrawn back into the vagina following dorsal wall and rectal penetration.

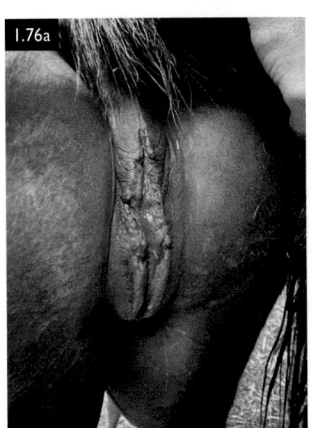 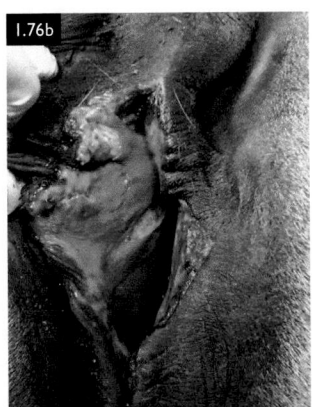 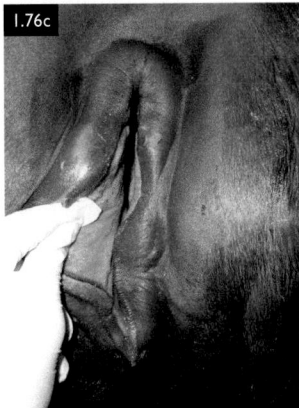

**FIG. 1.76** (a) first-degree perineal laceration; (b) second-degree perineal laceration; (c) third-degree perineal laceration.

## Diagnosis

- clinical signs.
- careful visual examination (speculum) and manual palpation of rectum and vagina.
- assess full reproductive tract for signs of bacterial or faecal contamination, especially in third-degree lacerations.
  - endometrial cytology/culture/sensitivity.

## Management

- early identification and correction of malpresentation during parturition will limit severity of injury.
- **first-degree laceration:**
  - often heal spontaneously:
    - daily warm water cleaning and antiseptic cream application.
  - surgical repair at 12 hours after foaling or 2–4 days to allow swelling to recede.
  - rebreeding at second oestrus unless wound healing is compromised.
- **second-degree laceration:**
  - generally, require surgical repair to re-establish the perineal body/vulva/vestibule.
  - only after all inflammation and swelling have subsided, which is facilitated by:
    - daily cleansing with warm water and application of antiseptic creams.
    - NSAIDs +/- antibiotics plus tetanus prophylaxis may be necessary.
  - healthy granulation tissue needs to be present for repair.
  - apply tail bandage, clean area and infiltrate with local anaesthetic:
    - sharp, careful debridement of wound edges.
    - anatomical layer repair to re-establish perineal body and vulvar/vestibular seals.
- **third-degree laceration:**
  - repair usually delayed for 4–6 weeks due to trauma, swelling, contamination etc.
  - initial treatment similar to second-degree lacerations but required for much longer (2–4 weeks).

- manual removal of faeces every 6–8 hours may also be required in some mares for varying periods depending on the extent and severity of the injury.
- staged removal of all dead and extensively damaged tissue by careful sharp debridement at an early stage will speed up the process of secondary intention healing.
- granulation, epithelialisation and healing of the damaged areas take between 4 and 8 weeks depending on the severity of the injury.
- surgical repair is undertaken once all the inflammation has resolved and is essential if return to breeding is required.
- mares with a foal at foot may have the surgery delayed until after weaning.
- pre- and postoperative measures to improve the success of the surgical repair are essential and must be carried out with great care:
- preoperatively:
  - decrease faecal volume/loosen consistency to ease defecation and decrease rectal impaction.
    - laxative diet.
    - decrease concentrates.
    - lush green grass.
    - pelleted grass rations.
  - laxatives including magnesium sulphate and bran mashes.
  - withhold all food 24 hours prior to surgery to reduce faecal surgical contamination.
- perioperatively:
  - medications 24 hours prior to surgery until 5–7 days post-surgery:
  - broad-spectrum antibiotics and NSAIDs
  - tetanus prophylaxis.
- surgical repair is undertaken in stocks under standing sedation, epidural anaesthesia +/- lidocaine infiltration.
  - many variations in the surgical procedure (undertaken in one or two stages)
  - basic principles of all of them should include:
    - use of strong absorbable monofilament suture material.

- minimal tension on the suture lines by careful and extensive dissection of tissue planes allowing apposition without tension.
- creation of a thick shelf between the rectum and vestibule.
- placement of all sutures with good bites of tissue to decrease breakdown of suture lines.
- ♦ exact surgical procedures are detailed in the standard surgical texts:
- ♦ one-stage:
  - vagina, perineal body and rectum all repaired at the same time.
- ♦ two-stage:
  - rectovaginal shelf constructed initially with perineal body rebuilt later.
- ○ preoperative diet and perioperative medications are continued postoperatively:
  - ♦ external perineal wounds are gently cleansed, and antiseptic ointment applied daily.
  - ♦ laxative diet is continued for 4 weeks.
  - ♦ monitoring for normal faecal passage is vital and any straining to pass faeces must be dealt with urgently.
  - ♦ external sutures are removed 10–14 days postoperatively.
  - ♦ endometrial cytology and culture performed at second oestrus period after the operation:
  - ♦ allows adequate time for healing and one oestrous cycle to help natural resolution of the preoperative contamination.
- ○ mare can be re-bred by AI 4 weeks post-surgery or 3 months for natural covering.
- ○ complications of the surgical procedure include total or partial wound breakdown, infection, urine pooling and constipation.
- ○ mare may be left with an incompetent anal sphincter and a permanent audible abnormality when exercising.

## Prognosis

- first and second-degree laceration future breeding prognosis is good.
- third-degree laceration prognosis is fair–guarded post repair.
- breeding prognosis for un-repaired lacerations vary:
  - ○ first-degree (good)
  - ○ second-degree (guarded)
  - ○ third-degree (poor).
- recurrence is not unknown for first and second degree but unusual for third degree.

# Rectovaginal/vestibular fistulae

## Definition/overview

- less common than perineal lacerations but similar aetiology.
- communicating hole between vagina and rectal cavity (rectovaginal fistula) or vestibule and rectal cavity (rectovestibular fistula):
  - ○ latter more likely to damage the perineal body.
- size can vary from <1 cm upwards but on average 5–10 cm:
  - ○ small holes may be suspected from uterine air/contamination (Fig 1.77).
- repair depends on size and location.

## Management

- avoidance, management and treatment is similar to third-degree perineal lacerations.
- repair is performed after a period of time to allow inflammation and swelling to subside and for a ring of granulation tissue to form:
  - ○ vaginal contamination with faeces may occur and regular daily removal is advised.
- small fistulae may heal spontaneously.
- large ones can reduce in size over 4–6 weeks, and this allows all inflammation to resolve.
- surgical preparation and postoperative care are similar to third-degree laceration repairs.
- different methods of surgical repair are described:

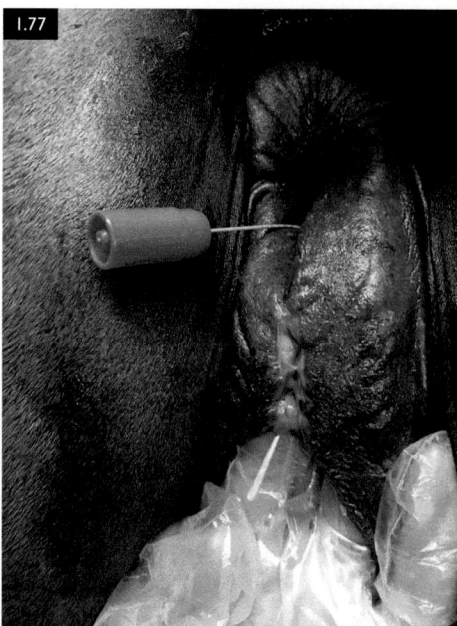

**FIG. 1.77** Small rectovestibular fistula <2 mm. Investigation of uterine air contamination lead to identification of a small fistula. The passage of a micro-tipped swab shows communication between vestibule and rectum. Transrectal repair was performed, and the mare subsequently conceived to both natural and AI breeding.

- fistula can be converted into a third-degree laceration then repaired:
  - useful if the perineal body is substantially damaged.
  - allows precise and strong internal repair.
- transverse perineal dissection to allow closure of the dorsal and ventral holes:
  - dissection route is then repaired.
- fistula visualised via a speculum either transrectally or transvaginally:
  - repaired with dissection, inverting and everting sutures.
  - limited exposure.
- Postoperative treatment and breeding advice is as per third-degree perineal lacerations:
  - rectal packing during natural covering to prevent excessive rectal pressure/damage can be helpful.
- complications include:
  - total or partial wound breakdown in rectal or vaginal layers.

- perineal dehiscence and secondary intention healing.
- perineal infection treated by drainage and lavage.
- uterine infection.

## Prognosis

- spontaneous healing carries a good prognosis.
- following surgical repair – depends on extent and severity:
  - fair–guarded for breeding.
- recurrence is uncommon.

## Post-foaling perineal bruising and vulval haematoma

### Definition/overview

- vulval/vaginal bruising is common, particularly in primiparous mares.
  - dystocia, manual manipulation and large foals can cause trauma.
- swelling, oedema, and bleeding often occurs:
  - perineal swelling may also occur and can be painful to palpate.
- often accompanied by vaginal or perineal lacerations (Figs. 1.78–1.80).
- complications can include severe vaginitis, fibrosis and abscessation.

### Management

- laxatives/faecal softeners to ease passage through swollen painful perineum/pelvic canal.
- NSAIDs, broad-spectrum antibiotics and tetanus prophylaxis may be necessary.
- daily cleaning and applying topical emollient creams.
- repair any lacerations after the swelling subsides.
- drainage of haematoma at 7–10 days is occasionally necessary.

### Prognosis

- good with most of the swelling resolving and damage repairing.

## Delayed uterine involution

### Definition/overview

- rapid involution occurs following parturition:

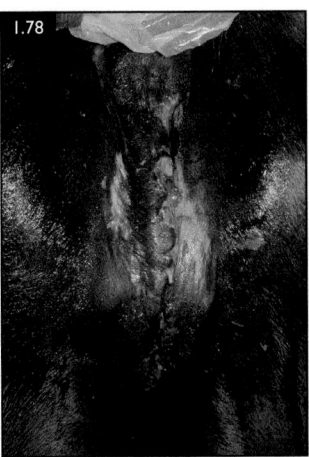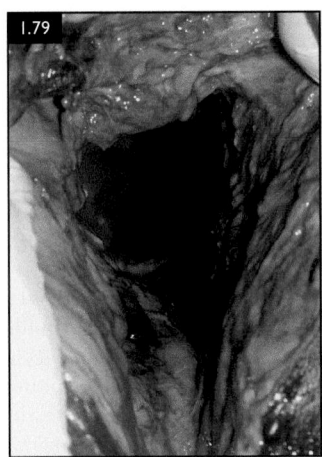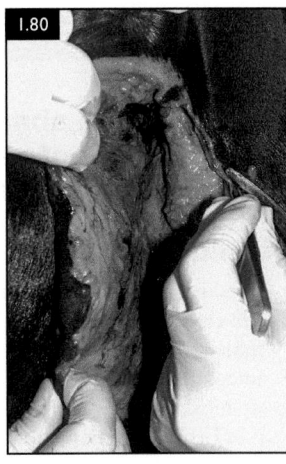

**FIGS. 1.78–1.80** (1.78) A Thoroughbred mare that suffered perineal, vulval and vaginal bruising and trauma, following major dystocia. (1.79) After retraction of the damaged vulval lips damage to the vestibule and vagina is obvious, while the cervix is still dilated and there is a view directly into the damaged uterus. (1.80) The vulval and vaginal damage is being lightly debrided with sharp scissors after thorough washing with antiseptic solution.

- o normal brown discharge (lochia) from the vulva for 3–7 days.
  - o uterus rapidly returns to pre-gravid size.
- histologically, the endometrium returns to a 'non-pregnant' state at 14–15 days post-partum.
- delayed involution where the normal process is interrupted due to a variety of reasons:
  - o uterus remains flaccid, voluminous and fluid-filled.
- urine pooling at cervix/within uterus may be additionally noted:
  - o mare's anatomy may contribute to gravitational retention of urine.
- financial/stud management impact as delayed breeding.

## Aetiology/pathophysiology

- most commonly occurs following:
  - o abortion
  - o dystocia and manipulation
  - o retained placenta     o placentitis
  - o uterine infection/haemorrhage
  - o lack of exercise.

## Clinical presentation

- often no clinical signs.
- occasionally prolonged vulval discharge post-partum.

- signs associated with uterine infection/fluid accumulation may be noted but are rare.

## Diagnosis

- ventral tilting of vagina on examination, with/without urine pooling and vaginitis.
- rectal palpation/ultrasound reveals distended uterus with hyperechoic luminal fluid.

## Management

- large-volume uterine lavage to remove inflammatory fluid.
- Oxytocin 10–20 IU i/m stimulates contraction and involution.
- paddock exercise.
- NSAIDs and antibiotics may be necessary if systemic signs are present.
- recovery from foaling can take time and some mares need to be rested for a season:
  - o perineal surgery (Pouret's procedure, urethral extension, uteropexy) may be required to prevent urine pooling and encourage uterine clearance.

## Prognosis

- good in mild cases which respond to early intervention:
  - o delay in breeding is often necessary.
- guarded to poor in chronic, unresponsive cases.

## DISORDERS AND DISEASES OF THE OVARY AND OVIDUCT

# Chromosomal abnormalities and disorders of sexual development of the mare

## Definition/overview

- horses have 64 chromosomes including 2 sex chromosomes (64XX female; 64XY male).
- uncommon conditions:
  o most common abnormality is 63XO (Turner syndrome in humans).
- most chromosomal abnormalities are unidentified unless bred or examined for breeding.
- chromosomal disorders are a developing area, but definitive diagnosis requires a karyotype.
- no treatments are available.

## Aetiology/pathophysiology

- usually associated with abnormalities in the sex chromosomes leading to:
  o abnormal female genitalia (internal and external) and infertility.
  o arises during embryogenesis or gamete formation.
- problem in mares bought for breeding, without veterinary pre-breeding examination.
- detailed internal examination (palpation/ultrasound) maybe required for identification.
- disorders of sexual development (DSD) include:
  o 63XO
  o mosaicism (2 different cell populations exist in the same animal):
    ♦ e.g. 64XX/63X and 63X/64XX/65XXX.
  o gene deletions.
  o SRY (sex determining region on the Y chromosome) abnormalities.
- all breeds are affected.

## Clinical presentation

- identified at breeding or pre-breeding examination.
- infertility     • subfertility
- abnormal oestrous cycles.

- repeated early pregnancy failure.
- mare may be physically smaller than normal or have conformational defects.
- external genitalia may be normal or abnormal e.g. small vulva or clitoral defects.
- internal genitalia abnormalities:
  o small/absent ovaries.
  o flaccid/absent uterus.
  o malformed cervix.
  o 'testicle-like' abdominal structures.
- occasionally seen in competition/race horses presented for abnormal hormone blood levels taken during competition.

## Differential diagnosis

- other causes of infertility.
- ovarian/endometrial atrophy in young (prepubertal) or old (senile) mares.
- post-exogenous hormonal administration.

## Diagnosis

- visual inspection of the external genitalia.
- rectal palpation and ultrasound of the internal reproductive organs:
  o small firm ovaries.
  o flaccid or absent uterus.
- cervical speculum examination may reveal cervical incompetence or abnormalities.
- Karyotype on a blood sample for definitive diagnosis (Figs. 1.81, 1.82).

## Management

- no treatment for reproductive dysfunction.
- occasionally source of hormonal abnormality (e.g. intra-abdominal testicles) may need to be removed.
- can be used as athletes/ridden horses.

# Ovarian tumours

# Granulosa (thecal) cell tumour (GCT)

## Definition/overview

- most common ovarian neoplasia.
- usually benign, slow growing, unilateral and vary in size at presentation (6–40cm):

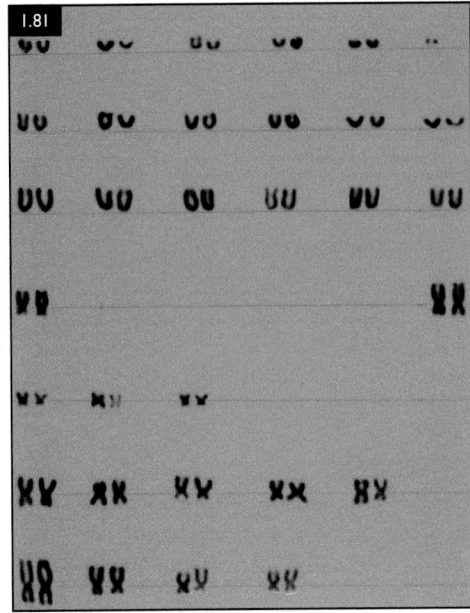

FIG. 1.81 Karyotype of a normal mare (64XX).

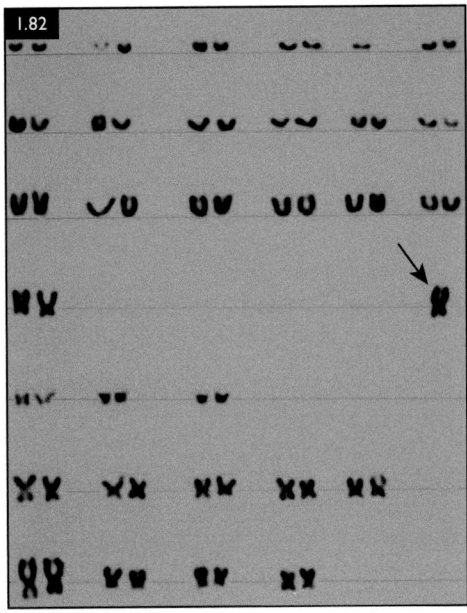

FIG. 1.82 Karyotype of a 63XO mare with a chromosomal abnormality. Note the single X chromosome (arrow).

- o characteristic 'honeycomb' appearance (Fig. 1.83).
- produces high hormone levels causing clinical signs and contralateral ovary atrophy.
- treatment is surgical removal.
- prognosis is good with most cases returning to normal cyclic ovarian function on the remaining ovary.

## Aetiology/pathophysiology

- tumour gradually replaces normal ovarian tissue.
- Inhibin (released by GCT) thought to suppress FSH levels and hence causes atrophy of contralateral ovary:
  - o rarely ovarian cyclicity continues but conception is rare.
- neoplastic cells can produce a variety of hormones:
  - o Testosterone – causes stallion-like behaviour.
  - o Oestrogens – causing nymphomaniacal behaviour and abnormal mammary development.
- hormone production can alter as the tumour grows and develops.
- Progesterone levels usually low.

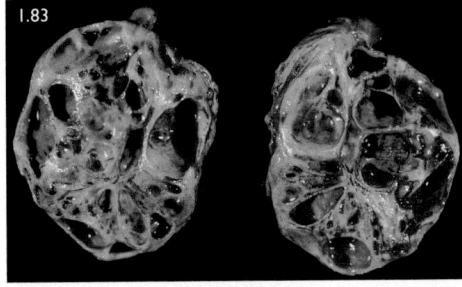

FIG. 1.83 Post-mortem specimen showing the cut section of a granulosa cell tumour of the ovary. Note the multiple honeycombed structures.

## Clinical presentation

- clinical signs depend on which hormone is released.
- stallion-like behaviour:
  - o aggression
  - o attraction to mares/mounting.
  - o chronically masculine muscle distribution
  - o occasional clitoral enlargement.
- constant oestrus or nymphomania.
- anoestrus or irregular cyclicity.
- rarely, unaffected ovary may continue to cycle, and tumours diagnosed in pregnant mares.

## Differential diagnosis

- Ovarian haematoma   • Cystadenoma.
- other ovarian neoplasias, including teratomas, lymphoma and lymphosarcoma.

## Diagnosis

- history and clinical signs are typical.
- palpation per rectum:
  - typically enlarged ovary with no palpable ovulation fossa and may be cystic.
  - unaffected ovary is usually small, firm and non-functional.
- ultrasonography (Fig. 1.84):
  - multicystic 'honeycomb' appearance.
  - varies from large single cyst to dense homogenous mass or haematomas.
  - contralateral ovary usually small, firm and inactive.
- hormone analysis:
  - ♦ raised Anti-müllerian hormone (>28.6 pmol/l [4 ng/ml])
  - ♦ Testosterone elevated in 50–60% cases (>0.17–0.35 nmol/l [50–100 pg/ml])
  - ♦ raised serum Inhibin (>700 ng/l [700 pg/ml])
  - ♦ Progesterone invariably low (<3.18 nmol/l [1ng/ml])
  - ♦ Oestogen rarely raised.
- histopathology of the removed ovary is definitive.

## Management

- surgical removal via flank or ventral midline laparotomy, colpotomy or laparoscopy:
  - latter is the approach of choice except where the ovary is very large.

## Prognosis

- good prognosis for return to fertility:
  - cyclic activity resuming within 9 months or the following breeding season.
- pregnant mares continue as normal to foaling but have abnormal cyclic activity post-partum.
- metastasis is rare.

# Other ovarian tumours

## Definition/overview

- includes teratoma, adenoma, cystadenoma, adenocarcinoma and dysgerminoma.
  - teratoma – second most common ovarian tumour.
  - all others are rare.
- non-secretory, unilateral and usually not malignant:
  - except for adenocarcinoma and dysgerminoma.

## Aetiology/pathophysiology

- **Adenoma:**
  - epithelial and usually unilateral.
  - usually arise from ovulatory fossa or ovulation fimbriae.
  - non-secretory, usually benign and can grow very large.
  - uni/multi lobular
  - contralateral ovary usually normal.
- **Cystadenoma** (Fig. 1.85):
  - cystic form of adenoma.
- **Adenocarcinoma:**
  - rare non-cystic form of adenoma.
  - metastatic.

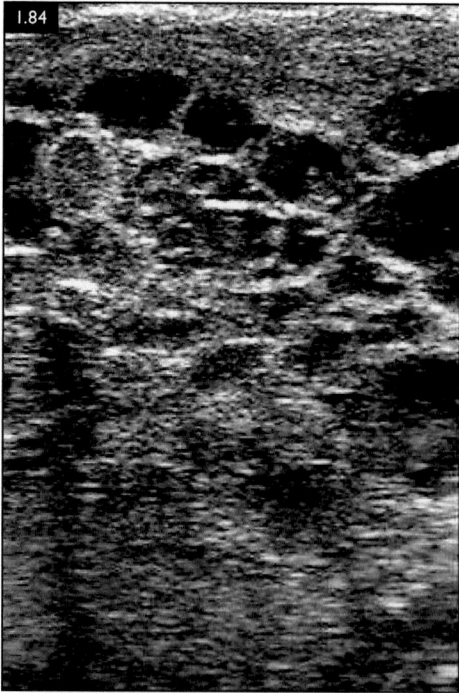

**FIG. 1.84** Transrectal ultrasonographic view of an ovary with a multicystic 'honeycomb' granulosa cell tumour.

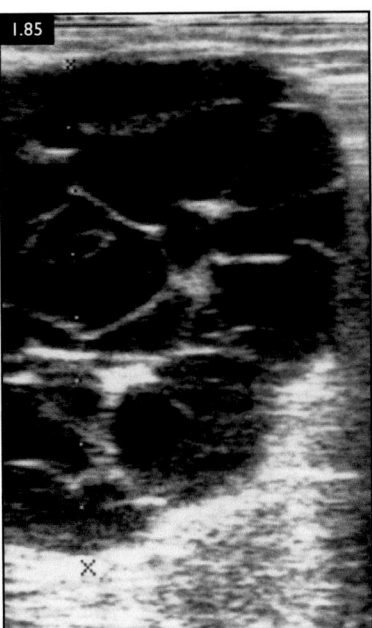

**FIG. 1.85** Transrectal ultrasonographic view of a left ovary with a cystadenoma.

- **Teratoma:**
  - solid or cystic tumour arising from germ cells.
  - benign and non-secretory.
  - contain abnormally placed embryonic structures e.g. hair, teeth, skin, nerves, bone.
- **Dysgerminoma:**
  - very rare and consists of homogenous, primordial germ-like cells.
  - lobulated or polycystic and non-secretory.
  - metastasises rapidly to thoracic and abdominal cavities.
  - associated with Hypertrophic osteopathy.

## Clinical presentation

- usually no hormonal/cyclic abnormalities as most are non-secretory:
  - contralateral ovary normal.
- unilateral large, irregular, abnormal ovary on rectal palpation and ultrasound.
- abdominal pain in large tumours due to traction on ovarian ligament:
  - rupture of ligament can cause intra-abdominal haemorrhage.

- fertility not usually affected unless anatomical impact from large tumour.
- metastasis may lead to abdominal pain, weight loss and/or ascites.

## Differential diagnosis

- Ovarian haematoma      o GCT
- lymphoma/lymphosarcoma.

## Diagnosis

- clinical signs, rectal palpation and ultrasonography:
  - appearance depends on type of tumour and tissue involved.
- hormone analysis often normal
- definitive diagnosis by histopathology following removal of ovary.

## Management

- surgical removal via laparoscopy (treatment of choice), colpotomy or laparotomy.

## Prognosis

- good in most cases which are benign and hormonally inactive:
  - poor for dysgerminomas and adenocarcinomas due to metastatic risk.

# Ovarian haematoma

## Definition/overview

- unilateral ovarian enlargement:
  - most likely cause of enlarged ovary during the breeding season.

## Aetiology/pathophysiology

- true haematoma forms from a *corpus haemorrhagicum* which greatly enlarges:
  - uncommon.
- Haemorrhagic anovulatory follicle (HAF) is not a true haematoma:
  - forms from intra-follicular bleeding.
  - ovulation generally does **not** occur (infertile cycle).
  - remains a feature on the ovary.
  - similar to Persistent anovulatory follicle (PAF) that does not bleed or ovulate.

## Clinical presentation

- detected during routine ultrasound examination.
- contralateral ovary remains normal and active with normal cyclic patterns.
- HAF/PAF form from apparently normal follicles:
  - do not grow on, ovulate or respond to ovulatory drugs as expected.

## Differential diagnosis

- GCT    • other ovarian neoplasia.
- Anovulatory follicle/transitional ovary
- Ovarian abscess.

## Diagnosis

- ultrasonography of ovary:
  - mass similar to a GCT.
  - HAFs usually:
    - more uniformly echogenic.
    - ovulation fossa present.
    - characteristic homogeneous appearance to their contents.
    - appearance changes more rapidly (within days) compared to GCT.
  - ultrasonographic appearance of PAF similar to a large follicle with hypoechoic fluid.
- HAF may luteinise and produce progesterone:
  - serum progesterone levels will rise.

## Management

- treatment not usually necessary as often resorbed.
- excessively large haematomas can be drained:
  - if interfering with normal reproductive function or causing pain.
  - via transvaginal or flank needle puncture.
  - removed via flank or midline laparotomy.
- prostaglandin to encourage luteolysis and organisation of haematoma:
  - **can increase incidence of HAF formation.**
- PAF and HAF most common in autumn and spring transition periods.

## Prognosis

- good unless the haematoma becomes very large.

## Oviductal disease

### Definition/overview

- lesions of the oviduct are rare:
  - blockage of the oviduct causing subfertility is recognised.

### Aetiology/pathophysiology

- uterine infection can spread to and affect the oviducts (salpingitis).
- inflammation of the oviduct can block/reduce patency thereby reducing fertility:
  - reduce passage of sperm, ovum or embryo.
- adhesions between the infundibulum and ovary/uterus are commonly found at post-mortem:
  - unknown significance on fertility.
  - may influence ability of fimbriae to received oocyte at ovulation.
- rarely affected by a tumour of the duct or ovarian cyst/tumour.
- as mare ages the debris, proteinaceous material and unfertilised oocytes accumulate in the oviduct slowing passage of sperm/embryo.

### Clinical presentation

- reduced fertility.
- endometritis.
- older mares (mean 18 yrs. old)

### Diagnosis

- full reproductive examination to rule out other common causes of infertility/abnormality.
- difficult to assess oviductal patency but test protocols are reported.
- ultrasound examination of the region:
  - assess any extra-oviductal lesions that may impact the oviducts.
- hysteroscopy to assess uterus, endometrium and uterine papillae.
- laparoscopic assessment.

### Management

- investigate and treat any endometritis:
  - bacterial culture/sensitivity
  - lavage    o infusions.
- adhesions can be broken down surgically.
- Prostaglandin E applied:

- directly to the oviduct via flank laparoscopy or:
- deep intrauterine deposition (hysteroscopic or blind).
- cause smooth muscle contractions and 'clearing out' of the accumulated debris.
- can be effective in the appropriate case.

- laser surgery via hysteroscopic approach.

## Prognosis

- not easily diagnosed or treated:
  - require extensive investigation to rule out other causes.
- guarded for fertility.

---

## DISEASES OF THE VULVA, VAGINA AND CERVIX

## Pneumovagina and increased contamination of the caudal reproductive tract

### Definition/overview

- loss of normal vulval/perineal anatomy through multifactorial causes including:
  - trauma
  - loss of perineal fat
  - age etc.
- results in compromise to anatomical protective barriers of caudal reproductive tract.
- leads to pneumovagina (air sucked into the vagina):
  - causes bacterial contamination of vagina, cervix and uterus.
  - inflammation and subfertility/ infertility.
- identification requires familiarity with normal anatomy.
- diagnosis by clinical and speculum examination
- bacterial swab/cytology.
- treatment with vulval and perineal surgery.

### Aetiology/pathophysiology

- vulva and vestibule act as first barrier between reproductive tract and the environment.
- position of the ischium/pelvic floor relative to the vulva and the vulval dorsal angle:
  - alters competence of the vulvovestibular seal.
  - predisposes mare to aspirating air and contamination into the reproductive tract.
  - normal mares have a vertical vulva with 80% below the level of the pelvic floor.

- some mares have greater percentage of the labia situated dorsal to the pelvic floor:
  - ◆ compromises vulval function (See Figs. 1.12, 1.13).
  - ◆ proportion appear to develop an appearance where:
    - anus and rectum are pulled cranially towards the caudal abdomen.
    - leads to the dorsal vulva becoming more horizontal and sloping.
    - allows air and faecal contamination to occur.
- vulval labia and constrictor muscles can be damaged by repeated foalings:
  - stretching and lacerations, as well as repeated vulval surgery.
  - damage may also occur to the perineal body and vestibular sphincter:
    - ◆ with, or without, vulval damage.
  - further compromises the functional barriers.
- other factors include:
  - poor inherited conformation.
  - ageing with loss of muscle tone in body generally and in caudal reproductive tract.
  - pendulous abdominal conformation and poor body condition.
- poor function of the caudal tract barriers leads to:
  - air and bacteria/debris entering the tract, especially at oestrus.
  - pneumovagina which desiccates and inflames the mucosa.
  - vaginitis/cervicitis that can ascend into the uterus and cause an endometritis.

## Clinical presentation

- often identified at a breeding soundness examination (BSE).
- history of infertility/subfertility/early embryonic loss/repeated endometritis.
- abnormal vulval conformation, structure and/or tone noted on visual and digital examination:
  - Caslick index helps determine the necessity for a Caslick's vulvoplasty operation:
    - length of vulva in cm × angle of declination of vulva.
    - <150 (preferably <100) associated with normal anatomy and better fertility rates.
    - score increases naturally with age.
- relative length/percentage of vulva above the ischium also determinant of need for surgery.
- part vulval lips manually above the ischium and note any aspiration of air into the vagina:
  - confirms the vestibulovaginal seal also compromised.
- may also be a history of aspiration of air during normal exercise (vaginal wind sucking).

## Differential diagnosis

- other causes of endometritis and infertility.

## Diagnosis

- clinical history and examination of external genitalia and perineum.
  - anatomy changes between oestrus and dioestrus.

- assessment should be carried out at the latter.
- perineal body assessed by placement of one finger into the rectum and thumb in the vestibule:
  - at least 3 cm of muscular tissue between the two.
- calculation of the Caslick index helps define if procedure is necessary.
- vaginoscopic examination may reveal:
  - air in the vagina.
  - vaginitis/cervicitis with congestion and mucus/discharge:
    - sometimes frothy and foamy due to air.
- swabs for bacterial culture/cytology confirm presence of endometritis.
- rectal palpation and ultrasound examination may reveal an air-filled vagina/uterus (Fig. 1.86) and cranial migration of the uterus into the abdomen.

## Management

- Caslick's vulvoplasty is the most effective treatment (Figs. 1.87–1.90):
  - essential the dorsal commissure is closed and kept closed.
  - mare will require a Caslick's operation for rest of her reproductive life (Fig. 1.91):
    - requires opening for breeding and foaling using local anaesthetic infiltration.
    - rapid repair thereafter to prevent reinfection.
- **every Caslick procedure must be done carefully with minimal tissue removal:**
  - occasionally scoring the mucosa is sufficient rather than removal.

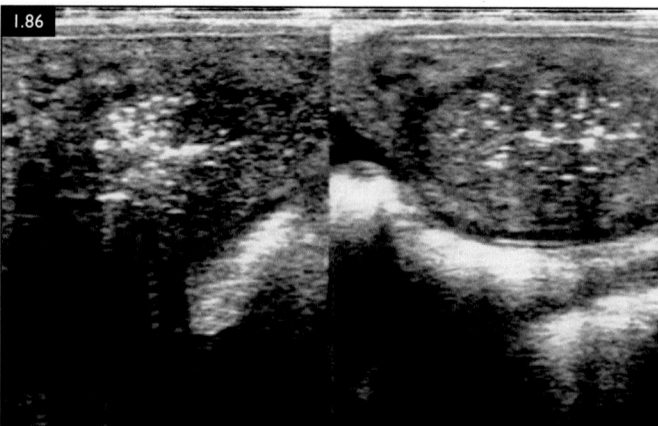

**FIG. 1.86** Transrectal transverse ultrasonogram of the right uterine horn showing air within the lumen (note the bright white hyperechoic foci). This mare had abnormalities of the vulva and vestibule leading to aspiration of air into the proximal reproductive tract.

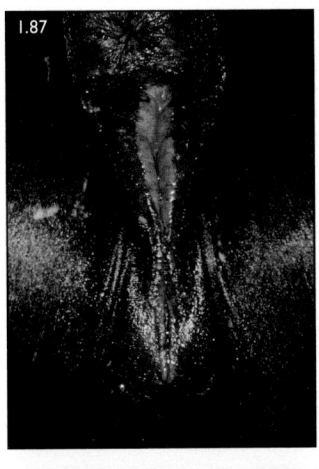

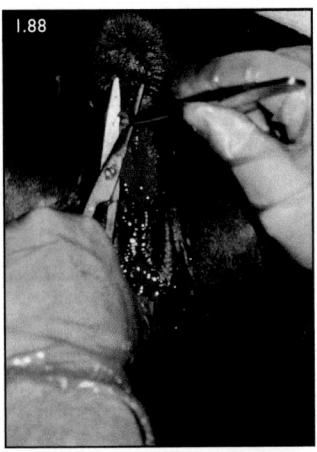

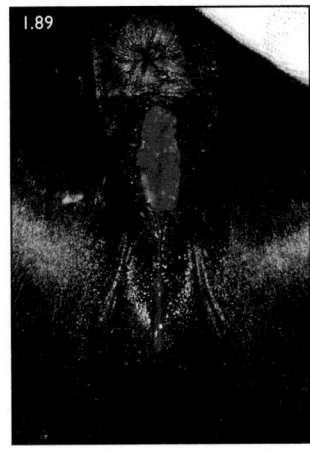

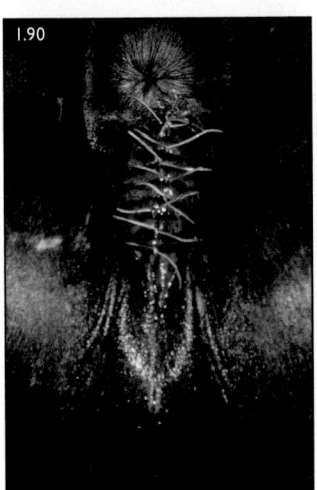

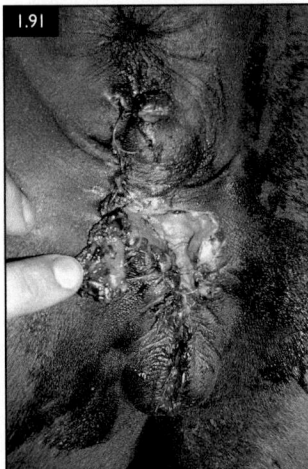

**FIGS. 1.87–1.90** A Caslick's operation. (1.87) Local anaesthetic has been infiltrated under the edges of the vulval lips down to the level of the pelvic floor. (1.88) Using sharp scissors, a thin slice of mucosa is removed just inside the vulval lips. (1.89) The two edges are now ready to be sutured together. (1.90) The operation is finished with the edges carefully brought together by simple interrupted nylon sutures.

**FIG. 1.91** A Caslick's operation undertaken post foaling that has subsequently broken down.

○ mare's breeding life can be shortened should excessive mucosal damage occur.
- endometritis/cervicitis/vaginitis secondary to the pneumovagina often resolve spontaneously after the operation due to natural defence mechanisms.
- older or heavily contaminated mares, may require specific endometritis therapy.
- any animals in poor condition will require appropriate feeding and care.
- Pouret's procedure is indicated where there is cranial migration of the anus and sloping of the vulva.
- Perineal reconstruction is indicated where there is vestibular/perineal insufficiency.

## Prognosis
- variable depending on severity of presentation.

- straightforward Caslick procedure which is well maintained carries a good prognosis.
- progression of the problem with repeated trauma or poor healing of the area can reduce fertility.
- perineal and vestibular insufficiency carry a slightly worse prognosis.

# Urovagina

## Definition/overview
- urine pooling in the cranial vagina:
  ○ conformational defects of the vestibule/vagina/perineal musculature.
- urine collects around the cervix leading to inflammation and uterine contamination:
  ○ chronic cases particularly can lead to infertility.

- may improve spontaneously, but surgical treatment is occasionally necessary.

## Aetiology/pathophysiology

- changes in conformation of the vulva/ vagina/vestibule resulting from:
  - age    ○ weight loss
  - foaling/mating injuries.
  - relaxation of the reproductive tract during oestrus.
  - Caslick's placed too ventrally below the level of the pelvic brim may interrupt urine exit.
  - large pendulous multiparous uterus/ abdomen can pull the reproductive tract cranially.
- may be a temporary state in the first post-partum oestrus while the uterus is still involuting.

## Clinical presentation

- vulval discharge.
- urine scalding around perineum.
- infertility.
- abnormal vulval conformation.

## Differential diagnosis

- other causes of infertility/subfertility
- endometritis   • vaginitis.
- Pneumovagina.

## Diagnosis

- urine and vaginitis at vaginoscopy at oestrus.

- manual examination.
- transrectal ultrasonography will identify signs of:
  - endometritis    ○ uterine fluid
  - poor involution.

## Management

- many acute cases, particularly post-partum, resolve spontaneously with uterine involution.
- sexual rest and improvement in body condition can be helpful.
- correct any excessively low Caslick vulvoplasty.
- vaginitis/cerviculitis/endometritis may require treatment:
  - lavage    ○ flushing
  - antibiotic infusion.
- severe or chronic cases often require surgical treatment:
  - urethral extension to move the urethral opening caudally (Fig. 1.92).
  - Perineal body transection (Pouret's procedure) (Fig. 1.93) may reduce urine pooling in some mares with abnormal anatomy.

## Prognosis

- good in the temporary condition in younger mares.
- fair to poor in older mares and chronic cases.
- mares may become intermittent breeders due to the recovery time needed after foaling.

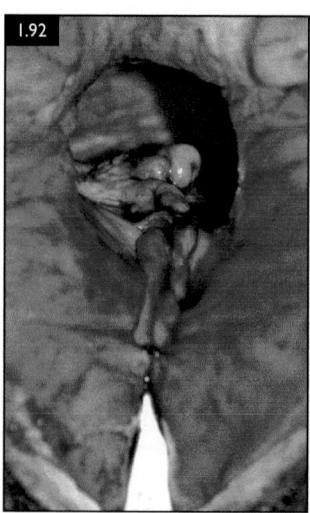

**FIG. 1.92** Immediately postoperative photograph of a completed urethral extension surgery with a bladder catheter in place to divert urine away from the surgery site. The vestibular fold is cranial and dorsal. Note how far caudally the catheter exits, which is where the urine will exit.

**FIG. 1.93** This mare has just had a Pouret operation carried out. Note the way the dorsal commissures of the vulval lips are vertical and there is a shelf effect above them before reaching the anus.

# Persistent hymen

## Definition/overview

- maiden mares often have remnants of the hymen present, of varying size and completeness.
- occasionally forms a complete membrane:
  - usually thin and easily broken down manually.
  - in rare cases, can be seen as a bulging membrane at the vulva.

## Aetiology/pathophysiology

- normal structure in the maiden mare:
  - cranial to urethral opening at junction of vestibule and vagina.
- usually, a thin structure forming a low ridge of tissue across the caudal vagina:
  - less commonly dorsally orientated strands or a complete membrane.
  - if complete, fluid from the endometrial glands may accumulate causing:
    - ◆ build-up of uterine/vaginal fluid.
    - ◆ rarely, bulging out of the membrane at the vulval lips.

## Clinical presentation

- if the hymen is present, may be broken down at the first covering or with manual vaginal examination, leading to minor bleeding.
- can prevent passage of speculum or hand, incorrectly suggesting a shortened vagina.
- bulging membrane (bluish-white) at the vulva in the young cycling mare, often seen at grass.

## Diagnosis

- vaginal examination – speculum or manual.
- rectal palpation/ultrasonography of the entire reproductive tract:
  - confirm normal anatomy cranial to the hymen (not a severe congenital deformity).
  - may be fluid accumulated cranial to the hymen in the vagina and, occasionally, in the uterus.

## Management

- simple manual vaginal palpation will breakdown strands and folds of membrane.
- membrane thick or complete:
  - sedate mare and carefully cut the centre with scissors.
  - then tear the membrane digitally through the new hole.
- accumulated fluid then drains away.

## Prognosis

- good with no long-term consequences in normal, uncomplicated cases.

# Varicose veins of the vagina

## Definition/overview

- quite common, particularly in older mares, and often noted in late pregnancy.
- prominent varicose veins in the mid-cranial either dorsal or ventral vaginal wall:
  - at the level of the hymen remnants.
- can cause intermittent vaginal bleeding and potential for secondary infection.

## Aetiology/pathophysiology

- cause is unknown, but they increase with age and multiparity:
  - suggests a response to hormone changes.
- generally, persistent once present.
- bleeding is usually minor and intermittent:
  - partial rupture of vessels usually in late pregnancy or occasionally, during oestrus.
- post-partum, bleeding usually stops:
  - possibly, less pressure/strain on the caudal reproductive tract and hormonal changes.

## Clinical presentation

- often incidental finding at reproductive examination.
- usually present as intermittent, slight, fresh haemorrhage from the vulval lips in:
  - older, late-pregnant mares or, less commonly, during oestrus, particularly after breeding.
  - once veins are established, bleeding may occur every year in late pregnancy or, in severe cases, at every mating.
  - increase in incidence and/or severity should be investigated further.
- bleeding is usually minor but can be more pronounced in rare cases.

## Differential diagnosis

- other causes of vulval discharge:
  - placental infection/inflammation.
  - urinary tract infection/inflammation.
  - trauma to the reproductive tract post breeding or examination.

## Diagnosis

- **thoroughly clean the perineum and use sterile lubricant and speculum:**
  - reduce iatrogenic introduction of bacteria.
  - **important when performing in the late-pregnant mare.**
- vaginal speculum examination will detect (Fig. 1.94):
  - enlarged (1–3 mm), thin-walled, bluish veins, usually dorsally at the level of the hymen.
  - severe cases may involve the ventral and lateral walls:
    - ◆ haemorrhage may be seen from the vessels or pooling in the ventral vagina.
- important to rule out other sources of bleeding from the reproductive or urinary tract, including the placenta:
  - rectal palpation, transrectal or transabdominal ultrasonography and endoscopy may be considered in certain cases.
  - ultrasound the caudal placental/reproductive tract in late pregnancy mare prior to introduction of speculum to assess other causes of bleeding first.

## Management

- treatment not necessary in incidental/intermittent cases.
- severe/persistent haemorrhage cases, vessel may be cauterised using diathermy/laser.
  - ligation of larger vessels is possible, but the walls are often friable.
  - injection of vessels with formalin solution/sclerosing agent under endoscopic control.
- Caslick's procedure to reduce air aspiration may be necessary if abnormal vulval anatomy.

## Prognosis

- variable depending on severity and need for treatment.

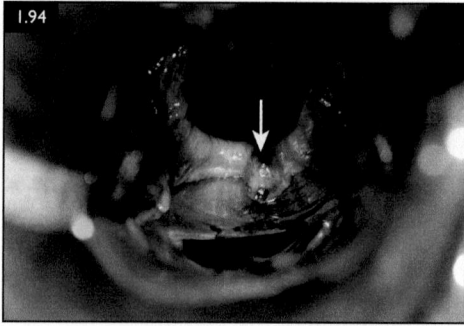

**FIG. 1.94** A view down a vaginal speculum of a case of vaginal varicosity showing the bluish enlarged vessel in the fold of tissue in the ventral vagina (arrow).

- with appropriate treatment, prognosis is fair–good.

# Cranial vaginal lacerations

## Definition/overview

- commonly related to accidents occurring at breeding, or occasionally during parturition.
- many occur dorsally and are easily missed.
- spontaneous healing is very quick in most cases.
- occasional complications, include eventration of bowel or bladder and/or peritonitis or local tissue necrosis following contamination.

## Aetiology/pathophysiology

- lacerations during parturition – usually caudal part of the vagina or cranial vestibule.
- cranial vaginal lacerations – usually associated with stallion's penis during breeding:
  - some stallions are more likely to cause these injuries.
  - likely to be related to a particular stallion's copulatory behaviour.
- breeding roll can be used to reduce trauma in maiden mares or with large stallions.

## Clinical presentation

- often, the laceration is not detected at breeding:

○ when the mare is examined vaginally again, it has healed.
- fresh blood may be noted on vulval lips/stallion's penis immediately after breeding.
- cranial dorsolateral vaginal wall:
  ○ often small (<5 cm) and partial thickness, with limited haemorrhage and rapid healing.
  ○ any contamination of the ejaculate with blood may affect its fertility.
  ○ full thickness lacerations are peritoneal:
    ♦ contamination by semen and penile/vaginal debris, and bacterial flora can lead to peritonitis.
    ♦ if noted early, mare monitored for signs of peritonitis for at least 3–4 days:
      – treatment instituted as soon as possible.
- severe vaginal lacerations or ruptures, particularly if they are ventral, can lead to bowel or urinary bladder eventration.

## Diagnosis

- clinical presentation and vaginal examination.
  ○ vaginal speculum examination:
    ♦ air may be aspirated when speculum inserted, suggesting peritoneal penetration.
    ♦ tear often cranial vaginal and dorsal to cervix.
- careful digital examination of the vagina to assess extent of trauma.
- paracentesis, abdominal ultrasound and haematology if peritonitis suspected.

## Management

- partial thickness and small lacerations:
  ○ often undiagnosed, left untreated and heal spontaneously without incident.
  ○ very few lacerations appear to develop into peritonitis.
  ○ mares can conceive from a breeding where a laceration has occurred:
    ♦ suggests ejaculation through the tear is rare.
- careful monitoring for complications for at least 96 hours:
  ○ prophylactic use of broad-spectrum antibiotics for 4–5 days.

○ usually all that is required for dorsal lacerations.
- ventral lacerations (rare) have an increased chance of eventration:
  ○ reduced by cross tying (discourage lying down), plus systemic medications to decrease abdominal straining.
    ♦ Caslick's procedure reduces air aspiration/contamination.
  ○ eventrated bowel or bladder should be protected and washed with sterile antibiotic solutions before replacement in the vagina and vulval suturing.
    ♦ epidural anaesthetic will stop straining.
    ♦ referral for surgical replacement and/or bowel resection.
    ♦ if unavailable, euthanasia should be considered.
- surgical repair of lacerations with absorbable sutures in the standing sedated mare under epidural anaesthesia can be performed in acute lacerations, although this is rare.
- some clinicians carry out a Caslick's vulvoplasty operation in cranial vaginal tears to reduce the possibility of air aspiration and contamination into the abdominal cavity.

## Prognosis

- small, partial thickness lacerations – fair to good.
- larger, full thickness lacerations – guarded.
- cases with eventration – poor.

# Coital exanthema

## Definition/overview

- sporadically occurring, usually venereally transmitted, infection of both sexes.
- caused by Equine Herpesvirus-3 (EHV-3).
- initial blisters, usually on the vulval lips, penis or prepuce, rupture to form ulcers.
- heal within 7–10 days if do not become secondarily infected.
- preventative applications of antiseptic creams and cessation of mating until healing occurs.

## Aetiology/pathophysiology

- highly infectious disease of external genitalia of mare and stallion caused by EHV-3.
- usually venereally transmitted:
  - other speculated transmitters include grooming and veterinary equipment.
- initial lesions on the mare or stallion 5–10 days after infected mating or contact with a contaminated fomite.
  - initially, blistering of the skin with vesicle formation occurs (1–3 mm diameter).
  - rapid progression to pustules, which rupture, leaving shallow ulcers.
  - rarely, lesions occur on the lips and nasal mucosa:
    - presence of respiratory disease is controversial.
- not known whether latent infections can occur with this specific Herpesvirus.

## Clinical presentation

- clinical lesions usually occur after a mating:
  - may develop spontaneously in mares after minor vulval injuries, damage or surgery.

- Mare:
  - lesions occur mainly on vulval lips and perineal skin:
    - occasionally on the anus and, rarely, in the vestibule.
- **Stallion:**
  - penis and prepuce are affected, particularly around the preputial fold.
  - pain on intromission and unwilling to breed.
  - pain on urination, especially if the urethral process has lesions.
- both sexes, early in the disease process, may develop swelling and pain in the affected areas.
- spontaneous healing occurs quite rapidly in 7–10 days in both sexes:
  - provided no secondary bacterial infection of the lesions occurs.
  - discharge from, and severity of, the lesions will increase if bacteria become established.
- after healing, the site of the lesions may be marked by permanent focal areas of lack of pigmentation (Figs. 1.95, 1.96).

## Diagnosis

- clinical signs.

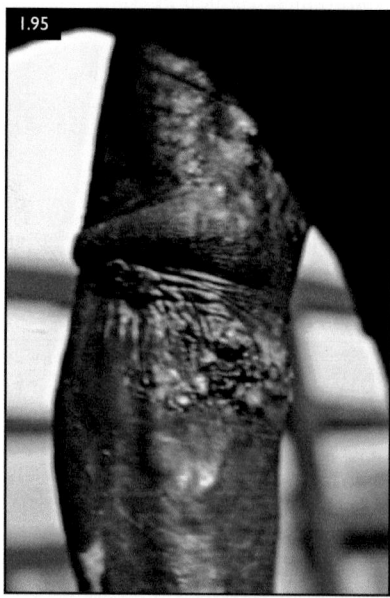

FIG. **1.95** Equine coital exanthema. This infection occurred in a 4-year-old first season stallion approximately 12 days post coitus with an infected mare.

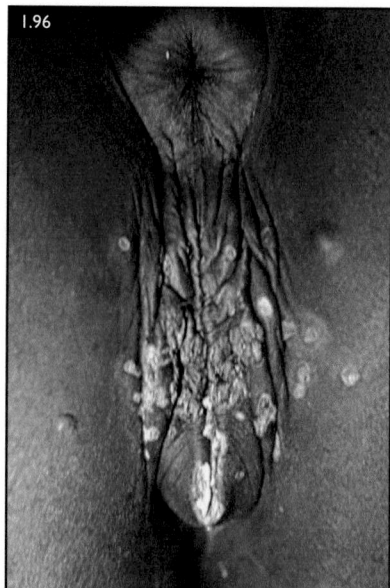

FIG. **1.96** Equine coital exanthema. Papules, pustules, crusts and erosion of the skin related to the vulva, anus and ventral surface of the tail have developed in this mare post service.

- EHV-3 isolation and intranuclear inclusion bodies on histopathology:
  - scrapes and biopsies of lesions.
- paired serology blood samples.

## Management

- prevent secondary bacterial infections symptomatically with antiseptic creams or sprays.
- other injuries to vulval tissue specifically treated.
- mating for stallion and mare should cease until all the lesions have healed.
  - mares can usually be mated on the next oestrus period.

## Prognosis

- no effect on fertility other than delay in breeding.
- immunity following natural infection is good although thought to be short-lived.
- no vaccine is available.

# Dourine

## Definition/overview

- venereally transmitted, protozoal disease present in certain parts of the world.
- clinical signs include fever, characteristic skin lesions, weight loss and discharges from, and swelling of, the external genitalia of the mare and stallion.
- organism is identifiable from discharges, blood and fine-needle aspirates of skin lesions.
- rarely treated and many countries have a slaughter eradication policy.

## Aetiology/pathophysiology

- caused by protozoal organism *Trypanosoma equiperdum*, transmitted venereally.
- present in Middle East, Africa, Central/ South America, parts of Asia and southeastern Europe.
- subject to national/international control measures to stop the spread of the disease.
  - complicated by the existence of clinically unaffected carriers.
  - epidemics reported where carrier or early infected animals are introduced.
- course of disease is slow, but progressive, and leads to high mortality rate (> 50%).

## Clinical presentation

- very slow onset with long incubation period (up to 26 weeks).
- clinical signs include:
  - fever.
  - urticarial-like skin plaques, particularly on the flanks.
  - muscle wastage and weight loss.
  - later on, neurological signs, collapse and death.
- **Stallion:**
  - early on mucoid/purulent urethral and/ or preputial discharge.
  - later scrotal and preputial oedema:
    - ◆ may extend cranially towards the ventral abdomen and chest.
- **Mare:**
  - early stages, mucoid or purulent vaginal discharge.
  - later, perineal and vulval oedema:
    - ◆ may extend cranially towards the mammary gland and ventral abdomen.

## Diagnosis

- history (including travelling), clinical signs and isolation of organism:
  - organism is present in low numbers but can be isolated from skin lesions, blood, CSF.
- serological testing important for international control.

## Management

- treatment is possible but rarely performed.
- eradication control measures including slaughter policies (country specific).

## Prognosis

- poor to grave.

# Tumours of the female reproductive tract (excluding ovary)

## Definition/overview

- uterine tumours are rare with Leiomyoma occasionally reported.
- vestibule/vulva are uncommon and include:
  - polyps
  - squamous cell carcinoma (SCC)

- melanoma in grey horses.
- mainly occur in older mares.
- biopsy and histopathology are essential for diagnosis.
- surgical removal may be appropriate in some cases.

## Aetiology/pathophysiology

- Leiomyoma – smooth muscle-like benign tumour of unknown cause.
- Polyps/papilloma thought to be caused by papillomavirus (benign).
- SCC more common in unpigmented skin, especially the perineal region (Fig. 1.97):
  - can be malignant and spread to local lymph nodes.
  - ulceration with local spread.
- Melanoma – very common in grey mares, especially in the perineal region:
  - may cause issues due to space occupation (usually benign) (Fig. 1.98).

## Clinical presentation

- Leiomyoma usually single nodules in uterine wall (2–15 cm):
  - usually found incidentally on rectal and ultrasound examination:
    - firm, spherical, hyperechoic masses in uterine wall.
  - occasionally cause slight bleeding if ulcerated.

- can cause subfertility/infertility.
- Polyps/papilloma usually protrude between vulval lips:
  - often pedunculated with little surrounding reaction.
  - slow growing but when larger may ulcerate and cause secondary infection.
- SCC may occur as two types:
  - proliferative forms occurring on vulva, vestibule and clitoral region (Fig. 1.97)
  - ulcerative forms occurring on perineum or vulval lips.
  - secondary infection is common leading to discharge and odour.
  - usually local spread, but occasionally to local lymph nodes and rarely beyond.
- Melanoma:
  - typical black lesion/s affecting the perineal region.
  - vary in size but can be very extensive.
  - may ulcerate with trauma or rapid growth:
    - predispose to local infection.
    - develop thick black haemorrhagic discharge.
  - can interfere with mating, AI or parturition.

## Diagnosis

- clinical history and gross appearance.
- vulval/vaginal examination.

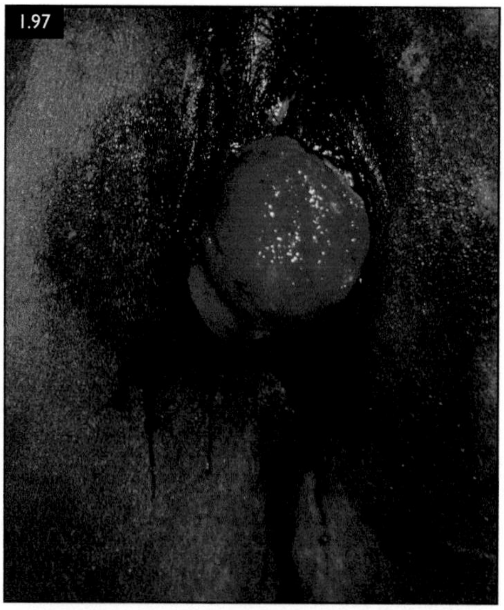

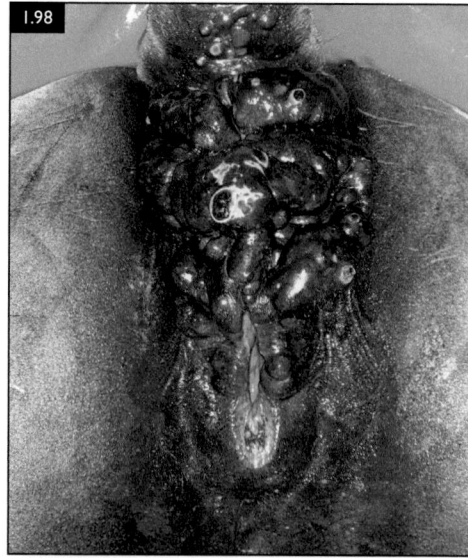

**FIG. 1.97** A squamous cell carcinoma of the clitoris and ventral vulval lips.

**FIG. 1.98** Massive infiltration of the perineum, tail head and dorsal vulva by a melanoma.

- rectal palpation/ultrasonography.
- biopsy and histopathology:
  - following excision or wedge/punch technique.
  - via hand- or endoscopically-guided if internal.
- palpate local lymph nodes in the inguinal and iliac regions.

## Management

- Leiomyoma – transendoscopic removal using loop diathermy or laser.
- Polyps/papilloma – ligated if pedunculated or surgically severed.
- SCCs:
  - surgical excision if small, accessible and localised.
  - other treatments if larger include:
    - cryotherapy.
    - local BCG or cytotoxic infiltration.
    - radiotherapy.

- Melanomas are rarely treated:
  - surgical excision if large or ulcerated.
  - intralesional injection of chemotoxic agents or vaccination may be appropriate.

## Prognosis

- Leiomyoma – can be difficult to remove totally and may affect fertility.
- Polyps/papilloma – good prognosis if easily and completely removed.
- SCCs – variable prognosis depending on location, type and extent:
  - worse for ulcerated or large lesions or where there is local spread.
- Melanoma – good prognosis in early stages:
  - large or extensive lesions with local infection:
    - may affect defecation, mating and parturition worsening the prognosis.

## DISEASES AND DISORDERS OF THE UTERUS

## Endometritis

- inflammation of the endometrium is a common cause of subfertility in the mare:
  - Persistent mating-induced endometritis.
  - Infectious endometritis.
  - Chronic degenerative endometritis.
  - Sexually transmitted diseases.
- uterus is subjected to numerous insults during the mare's life:
  - routine cyclic changes.
  - non-routine challenges:
    - mating    ◆ pregnancy
    - parturition  ◆ trauma to tissues.
    - bacteria/fungi/semen/faeces/urine contamination.
- well established uterine defence mechanisms including anatomical and immunological measures.
- any impairment of defence measures may compromise uterine function.
- age and multiparity alter the tract and may increase susceptibility to infection/ reduce effective response to infection.
- trauma and subsequent healing may alter anatomical defence mechanisms.

- failure of mechanical clearance of fluid, debris, and inflammatory substances from the uterus is a major predisposing factor to the development of endometritis.

## Persistent mating-induced endometritis (PMIE)

### Definition/overview

- mild, transient endometritis is considered normal following mating, foaling, examination or treatment with foreign substances.
- mare should be able to resolve the inflammation rapidly, usually within 12–24 hours.
- uterus does not evacuate fluid/recover within expected time:
  - prolonged uterine inflammation leads to reduced fertility.
  - increased EED.
  - primarily due to ineffective uterine contractility and lymphatic drainage.
- contributory factors include anatomical abnormalities:

- pendulous uterus in multiparous mares – gravity retains fluid within the uterus.
- poor cervical dilation in older maiden mares – fluid retention.
- persistent accumulation in uterine lumen of inflammatory products and debris leads to:
  - further inflammation, oedema and damage.
  - increasing likelihood of infections and further endometrial damage and fibrosis.

## Clinical presentation

- multiparous mares and older maiden mares (teenage):
  - poor perineal/vulval conformation leading to pneumovagina.
- cervical pathology in any aged mare – fibrosis or incompetence.
- often no signs prior to mating:
  - subsequent uterine luminal fluid retention within 12–36 hours following mating.
- early pregnancy loss with repeated early return to oestrus ('short cycling').

## Diagnosis

- post-breeding examination (rectal ultrasonography) 12–24 hours post mating shows:
  - intra-luminal uterine fluid accumulation (Figs. 1.99. 1.100).

- sampling of the fluid for cytology and bacterial culture/sensitivity:
  - neutrophils evident on cytology.

## Management

- early recognition is essential.
  - ideally identify the 'high-risk' mare prior to breeding:
    - age.
    - previous breeding history.
    - BSE, including anatomical assessment.
    - endometrial swabs prior to breeding to identify inflammation.
    - method of breeding – natural versus frozen semen versus chilled semen.
- pre-breeding options include:
  - uterine lavage     ○ oxytocin.
  - corticosteroid:
    - dexamethasone or prednisolone at time of breeding/12 hrs. prior.
    - down-regulates post-mating uterine immune response.
    - **caution with use re laminitis risk.**
- post-breeding options include:
  - uterine lavage with 1–3 litres warmed saline/lactated Ringer's solution:
    - specific uterine lavage fluids see Chronic uterine infection section.
    - no earlier than 4 hrs. after mating.
    - plus oxytocin – 10–20 IU i/v or i/m q4–6h.

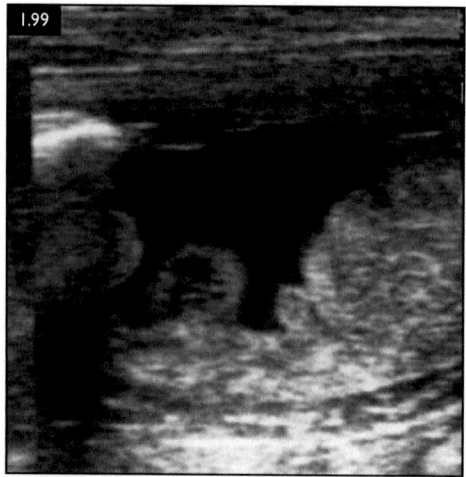 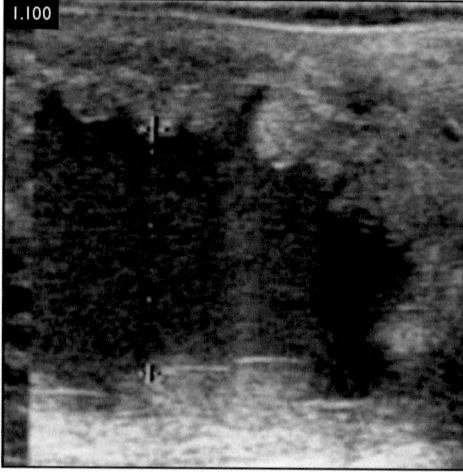

**FIGS. 1.99, 1.100** Transverse transrectal ultrasonograms of a uterine horn containing anechoic (1.99) and echogenic (1.100), more cellular, luminal fluid. Note the uterine wall oedema.

- ◆ hastens removal of post-coital debris, contamination and fluid without affecting the semen and before the mare mounts a major inflammatory response.
- ○ intrauterine antibiotics if indicated by culture/sensitivity.
- ○ treatments repeated at 12–24 hours or until no evidence of fluid remains.
- ○ cases of poor cervical relaxation:
  - ◆ Prostaglandin E directly onto the cervix + digital dilation can improve drainage.
- ○ Caslick's procedure if appropriate.
- ○ chronic and established endometritis:
  - ◆ active ongoing inflammation, severe oedema/lacunae or persistent luminal fluid.
  - ◆ sexual rest for two to three oestrous cycles may be warranted.

## Prognosis

- pre-breeding planning improves the prognosis for fertility.
- generally, decreases with age and with persistence of fluid.
- persistence despite treatment is a poor prognostic sign.

## Chronic uterine infection (CUI)

### Definition/overview

- may occur in PMIE mares if repeatedly bred in one season and not treated appropriately.
- repeated and untreated endometritis leads to:
  - ○ chronic endometrial changes, fibrosis and degeneration.
- commonly isolated pathogens include:
  - ○ β-haemolytic streptococci (*S. zooepidemicus*)
  - ○ *Staphylococcus aureus*.
  - ○ *Escherichia coli*
  - ○ *Pseudomonas* spp.   ○ *Klebsiella* spp.
  - ○ yeast and fungi (*Candida* spp, *Aspergillus* spp., *Mucor* spp.).
- predisposing factors include:
  - ○ pneumovagina
  - ○ incompetent cervix
  - ○ parturition.
  - ○ repeated mating

- ○ veterinary gynaecological examinations.

## Clinical presentation

- clinical history of:
  - ○ repeated breeding and poor fertility.
  - ○ PMIE.
  - ○ conformational abnormalities of the caudal reproductive tract.
  - ○ repeated intrauterine therapies.
- vulval discharge of varying nature and amount.

## Diagnosis

- clinical history.
- endometrial culture/sensitivity and cytology:
  - ○ neutrophils +/- bacterial growth.
- ultrasonography – uterine intraluminal fluid.
- endometrial biopsy (Fig. 1.101)

## Management

- initial approach involves:
  - ○ obtain endometrial swab or low volume lavage:
    - ◆ identify bacteria and sensitivity.
  - ○ under minimal-contamination techniques (see box):
    - ◆ uterine lavage – saline/lactated Ringer's solution for 2–3 days.
    - ◆ appropriate antibiotic infusion daily for 3–5 days.
  - ○ correct any perineal abnormality – Caslick's procedure.

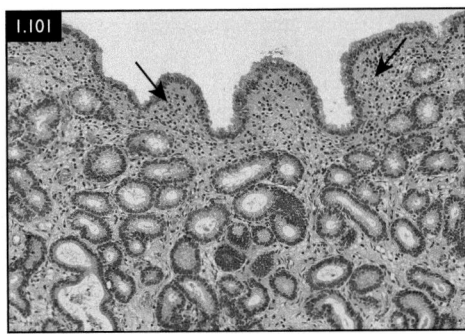

**FIG. 1.101** Histopathology of an endometrial biopsy from a mare with chronic infiltrative endometritis that subsequently responded to sexual rest and intrauterine therapy. Note the diffuse cellular inflammatory reaction (arrows).

- approach to a mare with CUI:
  - one approach does not work for all cases.
  - antibiotics guided by culture/sensitivity.
    - infusion volume 50–200 ml to ensure full endometrial coverage.
    - avoid irritant solutions and buffer appropriately as necessary.
  - uterine lavage:
    - fluids can be irritant (even lactated Ringer's solution (LRS)).
    - appropriate concentrations are needed (e.g. povidone–iodine).
    - some substances should be completely avoided (e.g. chlorohexidine gluconate).
    - minimal manipulation during lavage to reduce inflammation.
- specific uterine treatments and management of the susceptible mare (Table 1.4):
  - **Mucolytics:**
    - 30% DMSO (50–200 ml in 1 litre).
    - 20% Acetylcysteine – 6 g in 150 ml saline.
  - **Biofilm treatment:**
    - Tris-EDTA/Tricide – infuse 250–500 ml into uterus.
      - lavage 6–24 hrs later, then daily to clean uterus.
    - Hydrogen Peroxide – 60–120 ml of 1% solution.
      - lavage at 24 hrs. with lactated Ringer's solution.
    - DMSO up to 30% as a lavage fluid
    - Antimicrobial peptide mimic reactivates dormant *Streptococcal* infection.
  - **Fungal infection:**
    - often iatrogenic following repeated antibiotic infusions.
    - specific anti-fungal infusion or oral ketoconazole:
      - acetic acid lavage 2% (20 mls. vinegar in 1litre 0.9% saline).
      - 20% DMSO lavage.
      - 1-3% Hydrogen peroxide lavage.
      - 0.1% povodine iodine lavage.
  - **Immune modulation:**
    - autologous plasma - 60mls. infused 12-24 hours post covering.

- dexamethasone/prednisolone as described previously.
  - correct any perineal abnormality.
  - sexual rest until confirmed and stable improvement.

> **Minimal contamination technique:**
> - Thorough cleaning of the vulva and perineum.
> - Using a clean rectal glove and sterile lubricant introduce a sterile uterine catheter.
> - With minimal manipulation insert the catheter through the cervix.
> - Infuse the antibiotic solution.

## Prognosis

- guarded prognosis depending on:
  - causative organism (fungi/yeast are difficult to treat).
  - biopsy result.

# Sexually transmitted endometritis

## Definition/overview

- venereal transmission – natural mating and AI.
- some are notifiable in some countries.

## Aetiology/pathophysiology

- diseases include:
  - EVA.
  - Coital exanthema (EHV-3).
  - Dourine.
  - CEM.
  - *Pseudomonas aeruginosa*.
  - *Klebsiella pneumoniae* capsule type 1, 2 and 5.

## Diagnosis

- pre-breeding swabs (clitoral and cervix) and appropriate blood tests are used to identify the carrier/infected animal.
- some breed societies have mandatory pre-breeding testing.

## Management

- depends on the specific disease – see appropriate section.

**TABLE 1.4** Drugs and their doses used in treatment of uterine infections

| DRUG | DOSE | COMMENT |
|---|---|---|
| Amikacin sulphate | 1–2 g | Gram-negative organisms |
| Ampicillin | 1–3 g | Not good against anaerobes |
| Carbenicillin | 2–6 g | Broad spectrum including *Pseudomonas* spp. |
| Ceftiofur | 1 g | Very broad spectrum |
| Gentamicin sulphate | 0.5–2 g | Gram-negative organisms. Requires buffering |
| Neomycin sulphate | 3–4 g | Gram-negative organisms. Often used with penicillins |
| Oxytetracyclines | 1–5 g | Gram-positive organisms, especially streptococci |
| Penicillin (Na or K salt) | 5,000,000 IU | Gram-positive organisms, especially streptococci |
| Penicillin G | 3–6,000,000 IU | Gram-positive organisms, especially streptococci |
| Ticarcillin | 1–6 g | Broad spectrum, but poor against *Klebsiella* spp. |
| **Combinations** | | |
| Neomycin + penicillin G | 2 g/3,000,000 IU | |
| Gentamicin sulphate + penicillin G | 0.5–2 g/3–6,000,000 IU | |
| **Oral** | | |
| Enrofloxacin | 8 mg/kg | Accumulates and concentrates in endometrium |
| **Antimycotics** | | |
| Nystatin | 250,000–1,000,000 IU | Dissolve in 0.9% saline solution |
| Clotrimazole | 300–600 mg | Daily or every 2–3 days for 12 days |
| Vinegar | 2% solution | 20 ml wine vinegar in 1 litre saline |
| Povidone–iodine | 1–2% solution | Individual mares may be very sensitive to povidone– iodine, causing uterine pathology. Use with caution! |
| **Others** | | |
| Saline | Infuse 1 litre at a time | Infuse until recovery becomes clear |
| EDTA-Tris | 250–500 ml | Lavage 24 hours later with saline/lactated Ringer's solution. Biofilm reduction |
| Hydrogen peroxide | 60 ml of a 1% solution | Lavage 24 hours later. Treatment of fungal endometritis and biofilm reduction |
| Acetylcysteine | 6 g in 150 ml sterile saline | Mucolytic and treatment of uterine infection |
| DMSO | 50–200 ml in 1 litre of saline | Follow with saline lavage. Mucolytic and biofilm reduction |

Adapted from Perkins NR (1999) Equine reproductive pharmacology. *Vet Clin North Am Equine Pract 15(3):*687–704.

- 'freedom from infection' testing may be required.
- Clitoral infections can be treated using:
  - antibiotic cream or topical silver sulphathiazine.
  - surrounding perineal area and vagina should be treated as well.
  - recurrence is common.
  - Clitorectomy should be considered in recurrent cases.

## Prognosis

- depends on the infection.
- Clitoral infections carry a good prognosis but may require repeated treatment/surgery.

## Chronic degenerative endometrosis

### Definition/overview

- wide range of histopathological uterine changes, particularly found in older mares.

### Aetiology/pathophysiology

- degenerative and progressive in nature.
- no active inflammation:
  - lymphocytic, plasma cell and macrophage infiltration may be noted on histopathology.
- sign of uterine 'wear and tear' and ageing.
- individually specific.

### Clinical presentation

- older mare.
- difficulty in conceiving or early embryonic loss/pregnancy failure.

- dysmature foal born with normal gestational length.
- older maiden mares may also exhibit these changes and suffer from subfertility.

## Diagnosis

- Uterine biopsy changes include (Fig. 1.102):
  - fibrosis – diffuse, peri-glandular or peri-vascular.
  - lymphatic stasis
  - uterine sacculation.
  - transluminal adhesions.
  - glandular nests
  - glandular/lymphatic cysts.
- Hysteroscopy

## Management

- evidence-based treatments have not been described:
  - therapies based on inducing superficial endometrial inflammation and tissue loss.
  - attempt to 'rejuvenate' the uterus.
- some proposed treatments may result in irreversible damage:
  - care and full disclosure with the client are recommended.
  - mechanical uterine curettage
  - uterine lavage with caustic substances.
    - ◆ variable success.
    - ◆ intrauterine/cervical and vaginal adhesions.
    - ◆ only use if all other conventional therapies have been tried.
- re-biopsy 4 weeks after treatment to assess response.
- minimal-contamination techniques and specific post-breeding management is required for mares following treatment.

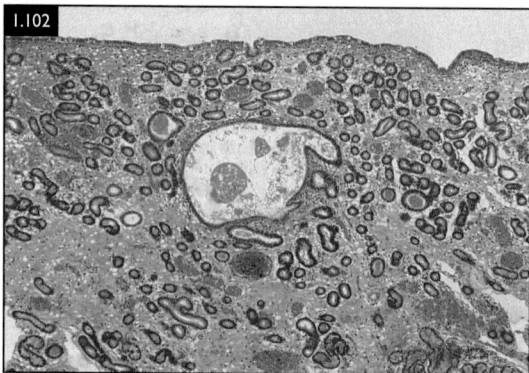

**FIG. 1.102** Histopathology of an endometrial biopsy from a mare with chronic endometrial degeneration showing a glandular cyst.

## Prognosis

- guarded to poor depending on the histological changes and response to therapy.

## Endometrial cysts

### Definition/overview

- endometrium contains a rich plexus of blood vessel and lymphatics.
- dilation and coalescence of lymphatic ducts results in:
  - thin-walled, fluid-filled structures.
  - within the endometrium or projecting into the uterine lumen.
  - uni- or multilocular.
- identification by rectal ultrasonography.
- clinical or breeding significance is debatable:
  - small, occasional cysts are of low significance.
  - large cysts, nests of cysts or cysts at the base of the uterine horn may affect fertility:
    - partial blockage of the uterine horn.
    - inhibits full movement of embryo during maternal recognition of pregnancy.
    - prevents attachment and placenta formation.
- treatment generally not needed but removal by loop diathermy or laser ablation is possible.

### Aetiology/pathophysiology

- fibrosis of endometrium due to trauma/age can cause lymphatic blockage and dilation.

### Clinical presentation

- often incidental on ultrasonographic examination of the uterus:
  - note during routine stud management by measuring cyst and documenting location.
  - cysts do not increase in size, move or develop an embryo inside.
- confusion with early pregnancy and potential twin pregnancy is possible.
- repeated early embryonic loss.

### Diagnosis

- fluid-filled (hypoechoic) structure on ultrasound examination (Fig. 1.103, 1.104)
  - variable size, usually <15 mm but can be larger.
  - usually 1–3, but multiple cysts can occur (termed a 'nest').
- hysteroscopy allows direct visualisation of the cyst.

### Management

- treatment is often not needed especially for small cysts in low numbers.

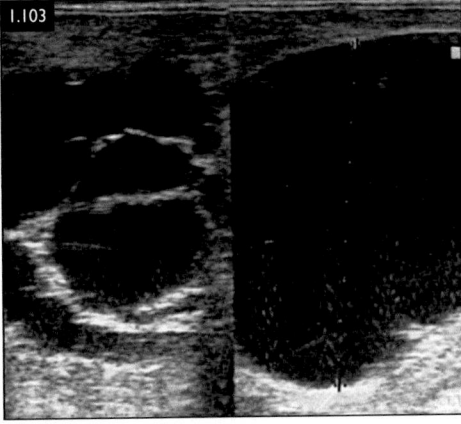

FIG. 1.103 Very large and multiple endometrial cysts seen in an aged mare on transrectal ultrasonography.

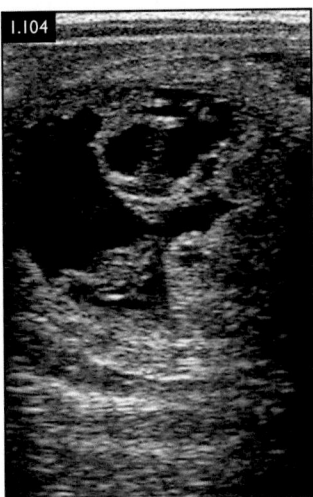

FIG 1.104 A 33-day pregnancy adjacent to a uterine cyst.

- treatment is indicated when:
  - large number of cysts present in a location that interferes with embryonic attachment/placenta formation.
  - confusion with early pregnancy is possible.
  - may cause the potential to miss a twin pregnancy.
- hysteroscopic laser ablation or removal via loop diathermy outside of the breeding season.
  - minimal scarring.

## Prognosis

- uncomplicated cases fair–good
- recurrence can occur especially where cyst is punctured rather than removed.

## Pyometra

### Definition/overview

- uncommon problem due to large volume of exudate accumulating in uterus.
- aetiopathogenesis is often unclear.
- often no systemic signs.
- can be difficult to treat and has a poor prognosis for future fertility.

### Aetiology/pathophysiology

- uncommon sequela to infection of uterus:
  - debris, inflammatory material and uterine secretions accumulate in uterus.
- mare often not cycling due to persistent CL.
- older mares or those with compromised uterine defence mechanism are more prone.
- rarely associated with an intrauterine foreign body or uterine neoplasia.

### Clinical presentation

- often no systemic illness.
- vaginal discharge may or may not be present:
  - when present often thick and purulent in oestrus.
- mare may or may not cycle.

## Differential diagnosis

- endometritis
- metritis
- pregnancy
- mummified foetus.

## Diagnosis

- rectal palpation and ultrasound will reveal:
  - enlarged, doughy uterus containing hyperechoic floccular fluid.
  - fluid volume can be significant (up to 60 litres).
- transabdominal ultrasound may be necessary to confirm the mare is not pregnant:
  - absence of a foetal body.
- blood analysis results are usually unremarkable.
- bacterial culture/cytology of the fluid useful but may be sterile.

## Management

- drainage via a sterile nasogastric/wide-bore Foley catheter or tube:
  - followed by repeated uterine lavage and fluid drainage until clean (may take days).
  - specific antimicrobial intrauterine therapy may be necessary.
- hysteroscopy/uterine biopsy post treatment to identify:
  - uterine adhesions.
  - underlying uterine abnormality and long-term prognosis for fertility.
    - ♦ may have severe irreparable endometrial damage or atrophy (risk during biopsy)
- risk of recurrence.
- ovariohysterectomy may be considered in severe cases (2 stage procedure):
  - difficult with risk of complications, especially cervical stump infection.

## Prognosis

- poor–hopeless for future breeding.
- mares that can still cycle and expel the exudate may respond to treatment.

## MISCELLANEOUS CONDITIONS

1

## Hermaphroditism/intersex

### Definition/overview

- genetic and phenotypic sex differ.
- Pseudohermaphrodites have ambiguous external genitalia.
- True hermaphrodites are rare.
- links to DSD (See page 70).

### Aetiology/pathophysiology

- congenital condition of unknown aetiology but possible causes include:
  - presence/absence of androgen receptors.
  - exposure of foetus to exogenous androgens.
  - genetic abnormalities.

### Clinical presentation

- wide variation including:
  - horses with normal female external genitalia:
    - masculine behaviour and testosterone.
    - male genotype (64XY).
    - often have short, blind-ending vagina
    - band/remnant of uterus and intra-abdominal testicles.

- horses with ambiguous external genitalia:
  - enlarged clitoral body that protrudes from the vulva.
  - short or long fused vulva.
  - female karyotype (64XX) (Figs. 1.105, 1.106)

### Diagnosis

- full reproductive examination including rectal palpation and ultrasonography.
  - **careful vaginal speculum examination:**
    - vagina may be shortened with a thin membrane between it and the abdomen.
- hormone analysis.
- chromosomal analysis (karyotyping).
- laparoscopy.

### Management

- may not be necessary.
- surgical removal of intra-abdominal gonadal tissue may be appropriate for athletic use:
  - laparoscopic approach is recommended.

### Prognosis

- horses are infertile.

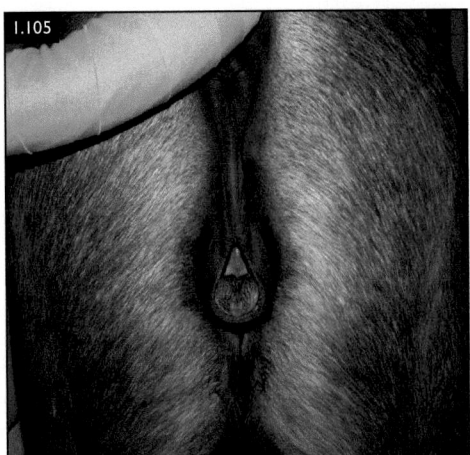

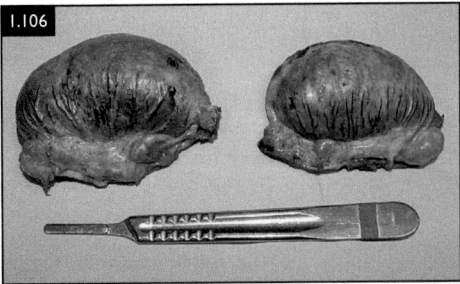

FIGS. 1.105, 1.106 The external genitalia (1.105) and removed gonads (1.106) of a true hermaphrodite. Note the enlarged clitoris and small vulval opening and removed ovotestes from the abdomen.

# Oestrous cycle anomalies and abnormalities

## Prolonged or persistent dioestrus

### Definition/overview

- common cyclic abnormality resulting from a persistent CL beyond usual 14–17 days.
- can persist up to 90 days:
  - must be differentiated from early pregnancy or causes of anoestrus.
- very effectively treated by intramuscular prostaglandin.
- good prognosis for a return to fertility.

### Aetiology/pathophysiology

- prolonged lifespan of CL in the absence of a pregnancy for up to 3 months.
- proposed reasons include:
  - inadequate release of prostaglandins (PG).
  - dioestrus ovulation leading to an immature CL:
    - not responsive to PG released in response to the original oestrus ovulation.
  - maternal recognition of pregnancy with subsequent early embryonic loss.
- occasionally associated with persistent uterine infection or pyometra decreasing PG release.

### Clinical presentation

- mare fails to exhibit normal oestrous cycles and oestrus behaviour in breeding season.

### Diagnosis

- rectal palpation and ultrasonography:
  - firm non-pregnant uterus.
  - active ovaries with follicular development.
  - presence of at least one CL (Fig. 1.107).
- differentiate from early pregnancy:
  - secretive breeding/mating is known in paddock bred horses.
- sequential blood sample analysis – persistent high levels of progesterone.

### Management

- intramuscular prostaglandin is treatment of choice:
  - CL over 5 days old will be lysed.
  - mare returning to oestrus in 5–7 days given normal follicular development is present.

### Prognosis

- good in uncomplicated cases.
- poor–hopeless in chronic uterine damage and pyometra cases.

## Behavioural anoestrus

### Definition/overview

- lack of behavioural oestrus despite normal oestrus physiological changes within the reproductive tract:
  - possible behavioural abnormality or immaturity.
- common in maiden mares or mares with foal at foot
- diagnosis confirmed by observing physiological signs of oestrus without behavioural oestrus.
- management by AI or careful and patient teasing to encourage mare to exhibit oestrus signs.

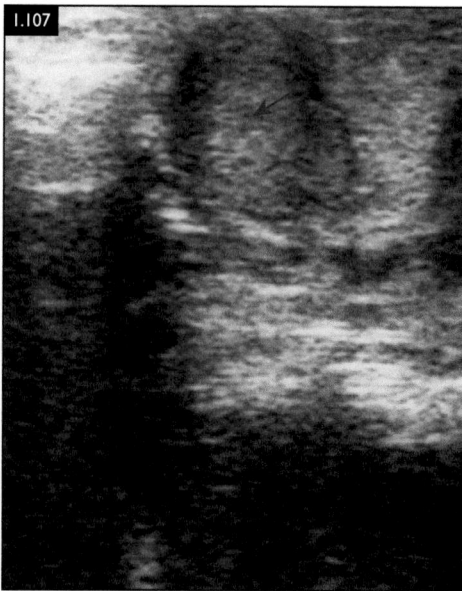

**FIG. 1.107** Transrectal ultrasonogram of the ovary of a mare in persistent dioestrus revealing a mature CL (arrow).

## Aetiology/pathophysiology

- usually, the hormonal, behavioural and physiological changes of the reproductive tract are coordinated:
  ○ in this condition, does not occur and the mare is not receptive to the stallion.
- may be overriding maternal factors that suppress behaviour or 'silent heat'.
- mares that have received anabolic steroids may display a similar condition:
  ○ this is iatrogenic and usually associated with little ovarian activity.

## Clinical presentation

- most common in maiden mares, mares with foals at foot, or nervous mares of any age.
- common when there is no access to teaser stallion or little effort is placed on teasing:
  ○ mares 'walked in' for covering.
- mare shows no oestrus behaviour and may be aggressive towards the stallion.

## Diagnosis

- repeated full reproductive examinations including rectal and ultrasonographic monitoring of the reproductive tract, and cervical assessment:
  ○ identifies physiological oestrus changes.
- serial plasma progesterone samples (every 4–5 days) should identify:
  ○ when mare enters oestrus (<3.18 nmol/l [1 ng/ml]).
    ♦ except for early in the year when winter anoestrus may still be present.
- careful and imaginative approach to teasing:
  ○ if necessary, restraint of the mare.
  ○ acute observation by teasing personnel, may allow recognition of oestrus behaviour.

## Management

- AI where this is permitted by the breed society:
  ○ timing requires veterinary monitoring as teasing behaviour is not reliable.
- careful and persistent teasing may improve the behaviour over time.

- care of the stallion and personnel at covering:
  ○ sedation of the mare +/- restraint may allow a safe covering.
- future oestrus periods may provide a better opportunity for covering.

## Prognosis

- fair with persistent and effective teasing.
- maiden mares may improve with patience and age.

## Post-partum or lactational anoestrus

### Definition/overview

- uncommon condition of the recently foaled mare where ovarian activity slows/ceases:
  ○ early foaling mares (reduced photoperiod).
  ○ poor or low body condition.
  ○ stress.
- mares may not have foal-heat ovulation or may shut down before the second oestrus period.
- often persists despite treatment for 4–12 weeks before restarting.
- diagnosis – full repeated reproductive examinations.
- prognosis – guarded for the current breeding season, can be repeated in subsequent years.

### Aetiology/pathophysiology

- multifactorial but related to photoperiod early in the year (January–March):
  ○ inadequate stimulation to maintain ovarian function in mares foaling early.
- coupled with nutritional needs and lactation demands from foal:
  ○ poor body condition and stress.
- may be related to persistence of the CL (prolonged dioestrus) after a normal foal heat or, more rarely, complete ovarian shut down and anoestrus after parturition.
- direct link to lactation is not proved.

### Clinical presentation

- commonly normal foal heat, both behaviourally and physiologically, and

then the mare does not return to normal cyclic behaviour at the normal interval.
- often foal heat is later than expected.
- rarely, mare goes straight into anoestrus post-partum.
- more common in mares that foal in the early months of the breeding season:
  - February/March in the northern hemisphere.
  - older mares in medium to poor body condition.
- may be 1–3 months before mares return to normal cyclic activity.

## Diagnosis

- full reproductive examination, including rectal palpation and ultrasonography:
  - assess ovaries and uterus.
  - follicular development behind expectations of the second oestrus period post-partum.
- plasma progesterone samples:
  - persistent CLs will have raised progesterone levels and relevant clinical signs.
  - true anoestrus mares – plasma progesterone concentrations <3.18 nmol/l (1 ng/ml):
    - small inactive ovaries, plus a flaccid uterus and partially open cervix.
- nutritional and/or medical conditions leading to poor condition/health require detailed investigation.

## Management

- early identification is essential if treatment is to be successful.
  - examination at 8–10 days post-partum (foal heat) for:
    - uterine assessment
    - bacterial swabs
    - ovarian activity.
- persistent CL responds well to intramuscular prostaglandin:
  - return quickly to a normal oestrous cycle.
- mare's environment and nutrition should be improved, and any stress factors removed.
- mares in true anoestrus, the treatment is often not effective:
  - similar regimes of management and drug therapy for the treatment of

transitional oestrus/anoestrus have been used with variable success in this condition.
- use of exogenous progestagens for 10 days is a good first-line treatment.
- experimentally, GnRH given via a continuous infusion pump delivering 2.5–5.0 µg/hour has led to cyclic activity within 10–14 days.
- GnRH analogue buserelin (12 µg q8h i/m for 500 kg horse) for 10–14 days, can be effective.
- oral domperidone (dopamine antagonist) at 1 mg/kg twice daily for 10–14 days can induce follicle activity and encourage lactation as a side-effect.
- prevention using extra light exposure (16 hours per day) for early foaling mares from early December to February/March (northern hemisphere) has been effective.

## Prognosis

- guarded in true anoestrus cases, particularly where the mare is stressed/poor condition.
- treatment failure is common as cases are often identified late in the process.
- some mares develop this condition repeatedly:
  - options are to breed every other year or to foster the foal.

# Dioestrus ovulation

## Definition/overview

- quite common for normal cycling mares to ovulate in dioestrus:
  - CL is present and functioning.
  - mare shows no oestrous behaviour.
  - only detected if the mare is examined regularly by rectal palpation and ultrasonography.
- normal variation in the oestrous cycle of the mare:
  - may be related to the mid-dioestral peak of FSH in the mare.
  - usually no effect on the oestrous cycle.
- when ovulation and luteinisation of a follicle occurs 1–4 days prior to prostaglandin release on day 15–16 of the cycle, the immature CL is unaffected and can be retained, leading to prolonged dioestrus.

- recognised from monitoring ovaries with ultrasound:
  - recent CL seen with older CL also present.
  - Doppler analysis reveals both are active.
- prostaglandin intramuscularly will lyse the new CL when it is responsive.
- no long-term effects on fertility.

## Failure of follicles to ovulate

### Definition/overview
- different to the Spring/Autumn transition phase anovulatory follicles.
- occurs during normal cyclical oestrus for unknown reasons.
- follicle does not ovulate, and mare remains in oestrus for prolonged periods.
  - persistence until regression or luteinisation occurs.
- diagnosis by repeated rectal and ultrasonographic examinations.
- treatment by ovulatory drugs but often unsuccessful.
- spontaneous resolution/regression leads to return of cyclical activity.
- recurrence is a problem within some mares.

### Aetiology/pathophysiology
- dominant follicle emerges but does not mature and ovulate:
  - some mares can have repeat anovulatory follicles over several cycles.
- Haemorrhagic and Persistent anovulatory follies have been described (HAF and PAF).
- unknown cause but likely to be endocrine related or due to abnormal follicular physiology.

### Clinical presentation
- mare may remain in oestrus for prolonged period of time.
- ultrasonographic monitoring shows the follicle persisting but not ovulating:
  - some may luteinise (hyperechoic appearance) and stop oestrus behaviour.
  - produce progesterone with a normal CL span or it may be prolonged.

- behavioural oestrus is usually normal, and mare will stand for covering:
  - no ovulation occurs so the cycle is infertile.

## Differential diagnosis
- ovarian tumours, including GCT.
- transitional ovaries.
- ovarian haematoma.
- ovarian abscess.

## Diagnosis
- repeated rectal palpation and ultrasound examination to identify and follow follicle/s:
  - ovary becomes enlarged on rectal palpation.
  - large follicle that does not progress to normal maturational changes towards ovulation.
    - feels firmer and within the ovarian structure.
    - large follicle (40 mm diameter) which may enlarge to >10 cm but does not ovulate.
- PAF have anechoic fluid whereas a HAF develops a variety of appearances (Fig. 1.108):
  - free-floating focal echogenicities.
  - bands of fibrin.
  - large haemorrhagic structures.

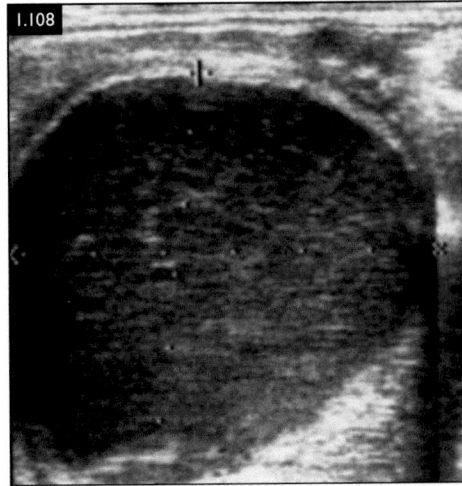

**FIG. 1.108** Transrectal ultrasonogram of the ovary of a mare that is failing to ovulate. Note the very enlarged anovulatory follicle containing echogenic material.

- contralateral ovary is normal anatomically and functionally:
  - compare with usual GCT presentation.
  - continue to cycle unless the HAF luteinises:
    - progesterone inhibits the return to oestrus.
- blood progesterone is raised if luteinisation of follicle.
- anovulatory follicles usually regress over a period of 1–2 months:
  - ovary returns to a normal size.

## Management

- many of these cases do not respond to treatment:
  - eventually spontaneously resolve the problem.
- if the follicle luteinises (HAF), may respond to exogenous prostaglandins:
  - use of PG may increase the risk of HAF formation:
    - particularly in early and late breeding season.
    - consider avoiding their use at this time.
- recurrence at subsequent oestrous periods can occur leading to:
  - wastage of covering, increased veterinary costs and semen wastage.
- treatment with hCG/GnRH is often unsuccessful as cases do not respond to endogenous LH.
- mating on spontaneous or induced ovulations in these cases are rarely successful:
  - oocyte ovulated is often degenerate and non-viable.

## Prognosis

- good for return to normal cyclical behaviour.
- short-term delays in breeding are an inevitable consequence.
- recurrence is common.

## Mastitis

### Definition/overview

- infrequent condition usually seen in the post-weaning mare, occasionally the lactating mare.
- variety of bacteria involved.

- swollen mammary gland which is hot and painful on palpation in acute cases:
  - secretions are often thick and clotted, occasionally with blood tinge.
  - harder and fibrotic in chronic cases.
- removal of the infected secretion, antibiotics, NSAIDs, and hot compresses are helpful.
- good prognosis in acute cases but guarded in chronic.

### Aetiology/pathophysiology

- usually seen in lactating mares or after weaning.
- most common in the summer and autumn months.
- bacteria involved:
  - *Streptococcus* spp., especially *S. equi zooepidemicus*, are commonly isolated.
  - range of gram-negative organisms and *Staphylococcus* spp.

### Clinical presentation

- mammary gland is hot, painful to palpation and swollen.
- ventral (unilateral) oedema, hindlimb stiffness and mammary vein enlargement in more severe cases.
- gland secretions can vary from watery to thick, clotted and occasionally blood tinged.
- systemic signs are occasionally present with pyrexia and inappetence.
- abscess formation (single/multiple) may occur with infection bursting out above teat.
- chronic cases become firm and fibrotic on palpation.

### Differential diagnosis

- mammary gland oedema
- abscessation
- trauma
- neoplasia.

### Diagnosis

- clinical signs and examination of gland secretions.
- cytology of secretions reveals large numbers of neutrophils with/without bacteria.
- bacterial culture/sensitivity is essential to direct therapy.

- ultrasound examination may help locate the site and morphology of any underlying abnormalities.

## Management

- drain the gland by stripping purulent material from the teats:
  - relieves pressure in the gland and improves mare comfort.
  - sedation may be necessary.
  - concurrent hot compresses.
- systemic antibiotics based on laboratory results:
  - therapy until 2–3 days post clinical resolution.
  - accompanied by intramammary versions.
    - often used in severe/refractory cases (off-licence cow formulations).
    - **infuse with care to avoid damaging teat canal.**
- systemic NSAIDs to decrease inflammation, pain and pyrexia.
- drain and flush abscesses, sometimes under GA.
- chronic mastitis is difficult to treat due to the fibrotic reaction that occurs:
  - mastectomy is necessary in some advanced and unresponsive cases.

## Prognosis

- acute cases often respond very favourably and quickly with little long-term damage.
- chronic mastitis carries a guarded to poor prognosis.

## Behavioural abnormalities associated with oestrous cycle

### Definition/overview

- mares are frequently presented for behaviour that owners link to female hormones ('moody mare')
- in-season behaviour is undesirable in some situations e.g. athletic/competition mares:
  - may be a perceived/actual change in the ability of the mare.
- direct link to the oestrous cycle is not as common as the presentation may suggest.
- important for owners to keep a diary over 4–6 weeks:

- document riding behaviour to provide a temporal link with the oestrous cycle.
- multiple treatment options are available.
- if behaviour is linked to oestrus, then good–fair prognosis.

## Aetiology/pathophysiology

- high oestrogen levels during oestrus (low progesterone) can cause behavioural changes.
  - behaviour is desirable in the breeding mare.
  - may be undesirable in the pleasure/ competition mare.
- behavioural changes are individually specific.
- high levels of progesterone (during dioestrus) may improve behaviour/ performance.

## Diagnosis

- actual temporal evidence is needed to link behaviour/performance to the cyclical activity of the ovary.
- ultrasound examination of reproductive tract when 'poor behaviour' is present to assess for signs of oestrus.
- hormone analysis is not always necessary.

## Management

- treatment only encouraged if there is a clear difference between oestrus and dioestrus behaviour.
- changing management/stabling may improve behaviour.
- medical treatment aims to maintain progestogen levels:
  - exogenous progestogen (some equestrian sports prohibit use of altrenogest).
  - maintain progestogen levels from persistent CL:
    - intrauterine PMMA sphere/self-assembling IUD:
      - placed within 24 hours of ovulation.
    - intrauterine plant oil:
      - placed on day 10 post ovulation.
  - oxytocin 25–60 IU q12h on days 7–14 post ovulation.

- other medical treatments:
  - GnRH agonist or GnRH vaccination.
  - Ovariectomy:
    - **caution as may not resolve the condition.**
    - allow to go into winter anoestrus to see effect of permanent surgery.

## Prognosis

- condition is complex and often multifactorial.
- where a temporal link is established then the prognosis is good:
  - otherwise, guarded to poor.
- treatments prolonging CL should last for approximately 90 days.

# Male Reproductive Tract

## ANATOMY AND PHYSIOLOGY

- neuroendocrine control of stallion reproduction involves the hypothalamus, pituitary gland, pineal gland, vomeronasal organ, and testes.
- stallion continues to produce sperm throughout the year, regardless of season.
- testicular size, semen production, libido, and hormone concentrations vary by season:
  - maximal values obtained in spring and summer months, lowest in the winter.
- hormonal control can be summarised:
  - GnRH is released in a pulsatile manner by the hypothalamus in response to both neural and hormonal control.
  - visual, tactile, auditory, and olfactory inputs are important regulators of stallion reproductive physiology:
    - exposure of stallions to mares increases GnRH and LH concentrations in the pituitary.
  - Flehmen response, or lip curl, exhibited by stallions investigating mares, is an important component of social interactions:
    - directs air across the openings of the vomeronasal glands to convey olfactory information from pheromones to the hypothalamus.
  - GnRH controls production and release of two gonadotropic hormones, LH and FSH:
    - act on cells of the testis, regulating spermatogenesis and steroidogenesis.
  - FSH regulates production by Sertoli cells of compounds important in sperm production:
    - androgen-binding protein, oestrogen, growth factors, inhibin, and activin.

- inhibin and activin feed back to the anterior pituitary to regulate FSH release.
- Sertoli cells regulate seminiferous tubular fluid, maintain the blood–testis barrier, and support the developing germ cells.
  - LH regulates the Leydig cells of the testis.
    - stimulates production of the steroid hormones:
      - testosterone, oestrogen and dihydrotestosterone.
    - normal testosterone levels locally within the testis are essential for normal spermatogenesis.
    - steroid hormones also regulate accessory gland function and maintain libido by systemic actions via the bloodstream.
    - testosterone and oestrogen feed back on the hypothalamus and anterior pituitary gland to regulate LH release.

## Testicular descent

- testicles normally descend into a scrotal position between:
  - last 30 days of gestation and first 10 days post-partum.
- in some colts, the testes may descend into the inguinal region and remain there for some time before fully descending into the scrotum.
- descent is coordinated by hormonal and physical events, including:
  - androgen production by the developing foetal gonads.
  - müllerian inhibiting and epidermal growth factors.

DOI: 10.1201/0781003386834-2

- o traction of the gubernaculum and elongation of the vaginal process.
- o intra-abdominal pressure.
- o expansion of the inguinal ring.
- homozygous deletion of the AKR1C gene was found in 6/67 (9%) of cryptorchid cases.

## Scrotum

- much less pendulous than in the ruminant species with two distinct scrotal pouches.
  - o protect, and thermoregulate the testes, epididymides, spermatic cords, and cremaster muscles.
- wall of the scrotum consists of four layers:
  - o scrotal skin is thin, generally hairless, and slightly oily.
    - ♦ contains numerous sebaceous and sweat glands, which assist in testis thermoregulation.
  - o tunica dartos layer lines both scrotal pouches and extends into the median septum, seen externally as the median raphae of the scrotum.
    - ♦ contraction or relaxation of this layer allows alterations in the size, shape, and position of the scrotum in relation to the body wall.
    - ♦ aiding testis thermoregulation.
  - o scrotal fascia is a loose connective tissue layer between the tunica dartos and the parietal vaginal tunic.
    - ♦ allows the testes and parietal tunic layer to move freely within the scrotum.
  - o parietal vaginal tunic is an evagination of the parietal peritoneum through the inguinal rings.
    - ♦ forms a sac that lines the scrotum.
    - ♦ apposed to the visceral vaginal tunic, which is the outer layer of the testis.
- vaginal cavity is the space between the parietal and visceral layers of the vaginal tunic.
  - o normally contains a very small amount of viscous fluid, allowing some free movement of the testis within it.
  - o fluid may accumulate within it as a result of a variety of causes.
- scrotum of the normal stallion should appear slightly pendulous, globular, and generally symmetrical.

- o skin should have no evidence of trauma, scarring, or skin lesions.
  - ♦ any lesions can cause significant alterations in testis temperature and affect fertility.
- o palpation of the scrotum reveals a thin and pliable covering, which slides loosely and easily over the contents.

## Testicles

- normal stallion testicles are palpable as two oval structures lying horizontally within the scrotal pouches.
  - o normal orientation of the testicle is ascertained by palpation of the tail of the epididymis and the ligament of the tail at the caudal pole of the testicle (Fig. 2.1).
  - o ligament is palpable as a fibrous nodule attaching the tail of the epididymis to the caudal pole of the testicle.
- examination of a normal stallion may identify rotation of up to 180° of one or both testicles.
  - o usually transient and non-painful.
  - o subsequent examination may reveal the testicle in normal orientation.
  - o must be differentiated from testicular or spermatic cord torsion.

## Epididymides and excurrent duct system

- each epididymis is a highly convoluted but unbranched, duct:
  - o approximately 70 metres long.
  - o grossly distinct head, body, and tail.
- head of the epididymis is a flattened structure that lies dorsomedially along the cranial border of the testis.
- body, or corpus, lies along the dorsolateral aspect of each testis.
- continues as the tail, a large, prominent structure attached to the caudal pole of the testis.
- during transport along the epididymis, the sperm undergo a number of morphological and physiological changes.
  - o ultimately renders them motile and fertile.
  - o specific changes include the shedding of the cytoplasmic droplet, plasma and

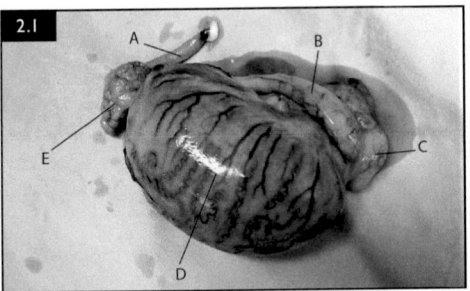

**FIG. 2.1** Gross appearance of the right testis and epididymis of the stallion. The head of the epididymis lies at the cranial pole of the testis (red arrow). The body of the epididymis courses dorsolaterally to the testis, and the tail of the epididymis lies at the caudal pole. A = deferent duct; B = body od epididymis; C = epididymis head; D = right testis; E = epididymis tail.

acrosomal membrane alterations, DNA stabilisation, and metabolic changes.
- o matured sperm is generally stored in the tail of the epididymis.
- deferent duct is the excretory duct for sperm:
  - o attaches to the tail of the corresponding epididymis.
  - o runs along the medial aspect of the testis.
  - o ascends via the spermatic cord through the vaginal ring into the pelvis.
  - o each deferent duct widens into its corresponding ampullary gland and

eventually terminates at the colliculus seminalis of the pelvic urethra.
- o colliculus seminalis is a rounded prominence situated on the dorsomedial wall of the urethra.
- o about 5 cm caudal to the urethral opening from the bladder.
- o site at which the ducts of the accessory sex glands empty into the urethra.

# Spermatic cord and vascular supply to the testis (Fig. 2.2)

- each spermatic cord is enveloped in the parietal layer of the vaginal tunic which extends distally from the internal inguinal ring and contains:
  - o deferent duct, testicular artery, testicular veins, lymphatic vessels, and nerves.
  - o cremaster muscle is situated in the caudolateral borders of each cord.
- tortuous testicular artery descends through the inguinal ring into cranial border of spermatic cord:
  - o divides near the testis into several branches to supply the testis and epididymis.
- corresponding network of veins leaves the testis and surrounds the testicular artery in a tortuous manner, forming the pampiniform plexus:
  - o cooler venous blood surrounding the testicular artery transfers heat away

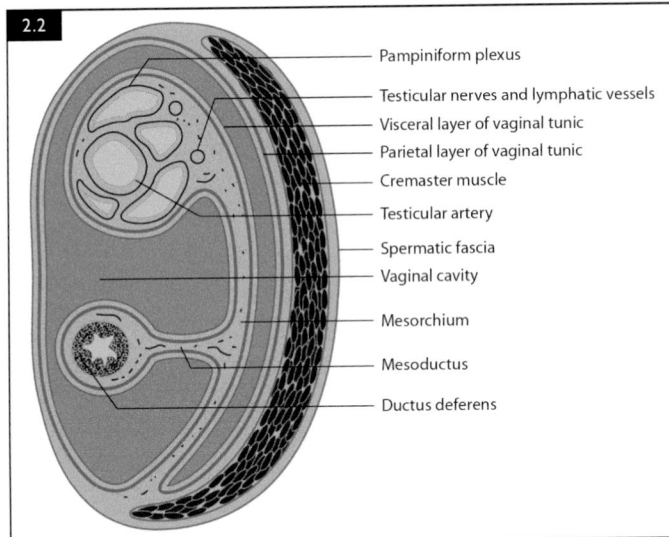

**FIG. 2.2** Schematic representation of a transverse section of the spermatic cord.

from the testicular arterial blood to the venous side.

- ○ responsible for much of the thermoregulation of the testis in the stallion.
- ○ also, evaporative heat loss through the scrotal skin.
- ○ blood within the testicular artery is several degrees cooler on reaching the testicle.

## Penis and prepuce (Fig. 2.3)

- penis of the stallion is musculocavernous in type and can be divided anatomically into:
  - ○ root, body, and glans penis.
  - ○ supported at its root by the paired suspensory ligaments of the penis and the ischiocavernosus muscles.
- cavernous spaces making up the erectile tissue of the penis are:
  - ○ single and dorsal corpus cavernosum, the corpus spongiosum, and the corpus spongiosum glandis.
  - ○ engorgement of these spaces with blood from branches of the internal and external pudendal arteries and the obturator arteries is responsible for erection.
  - ○ cavernous spaces within the penis are continuous with the draining veins.
- corpus spongiosum originates at the pelvis as the bulb of the penis and distally surrounds the penile urethra within a groove on the ventral side of the penis.

- ○ continues distally over the free end of the penis to form the glans penis (corpus spongiosum glandis).
- ○ responsible for the distinct bell shape of the stallion's penis during full erection.
- urethral process is distinctly visible at the centre of the glans penis:
  - ○ surrounded by an invagination known as the fossa glandis.
- ventral to the urethra and along the entire length of the penis is the bulbospongiosus muscle:
  - ○ smooth rhythmic contractions of the bulbospongiosus muscle assist in moving the penile urethral contents (semen and urine) distally.
  - ○ these pulsations are distinctly felt during ejaculation if a hand is placed on the ventral aspect of the penis during collection/natural service.
- paired retractor penis smooth muscles run ventrally along the length of the penis and attach at the glans penis:
  - ○ function to return the penis to the sheath following detumescence.
- prepuce is formed by a double fold of skin and resembles scrotal skin:
  - ○ essentially hairless and well supplied with sebaceous and sweat glands.
  - ○ external part of the prepuce, or sheath extends cranially from the scrotum before reflecting dorsocaudad to the abdominal wall to form the preputial orifice.

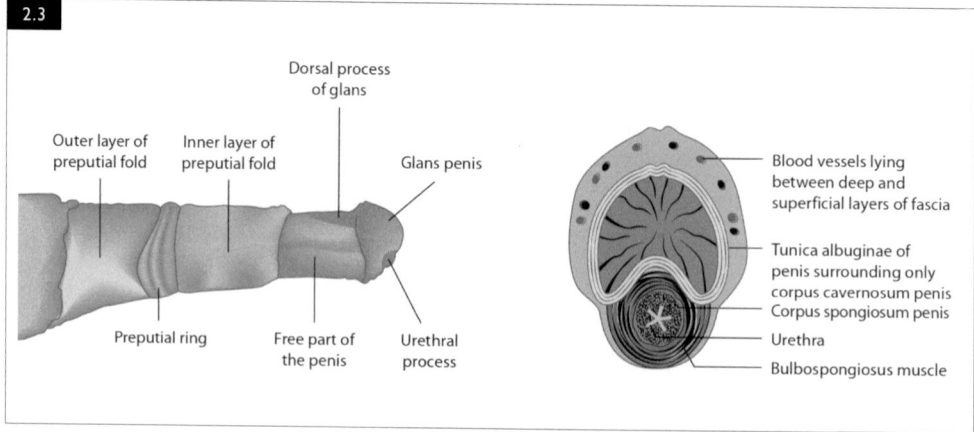

**FIG. 2.3** Schematic diagram of the erect penis and prepuce of a horse (left) and a cross-section of the free part of the penis (right).

○ internal layer of the prepuce extends caudad from the orifice to line the internal side of the sheath:
  ♦ reflects craniad towards the orifice again before reflecting caudad to form the internal preputial fold and preputial ring.
  ♦ additional internal fold allows the marked lengthening (approximately by 50%) of the stallion's penis during erection.
○ during erection:
  ♦ preputial orifice is visible at the base of the penis just in front of the scrotum.
  ♦ preputial ring is visible approximately mid shaft in the penis.
  ♦ distal to the preputial ring is the internal layer of the internal preputial fold.
• **newborn colts are nearly always born with the internal fold of the prepuce attached to the glans penis:**
  ○ within first 10 days of life the prepuce separates from the glans.
  ○ allows the prepuce to unfold and the penis to drop.
  ○ urethral process is centred to the preputial ring allowing urination to occur.
• penis and prepuce of a breeding stallion are best examined following teasing with an oestrus mare.
  ○ stallion can be observed to drop the penis and attain a full erection.

○ removal of smegma accumulations may be required for a complete examination of the surfaces.

## Accessory sex glands

• accessory glands found in the stallion are the prostate gland, seminal vesicles, bulbourethral glands, and ampullae (Fig. 2.4).
• secretions produce the seminal plasma, forms the majority of the ejaculate volume.

### Prostate gland

• central isthmus and two lateral lobes that extend along the caudolateral borders of each vesicular gland.
• lobulated or nodular and firm:
  ○ distinguishing it from the smooth, thin-walled vesicular glands lying next to it.
• each prostatic lobe measures approximately 5–9 cm long, 2–6 cm wide, and 1–2 cm thick.
• easily identified ultrasonographically:
  ○ two distinct symmetrical, homogeneously echogenic lobes (Fig. 2.5).
  ○ hypoechoic dilations within the gland parenchyma of each lobe usually evident:
    ♦ vary in size depending on frequency of ejaculation and degree of sexual stimulation.

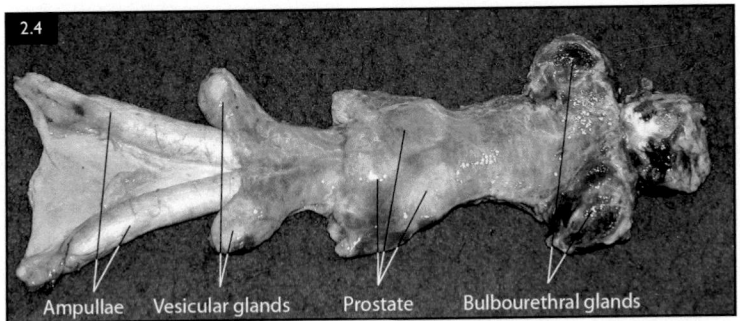

FIG. 2.4 Accessory glands of the stallion removed en bloc with cranial to the left. The ampullary glands are the most cranial of the accessory glands and are glandular dilations of the deferent ducts. The paired vesicular glands are found on either side of midline, just caudal to the ampullae. The prostate gland lies just caudal to the vesicular glands and includes a central isthmus and two lateral lobes. The paired bulbourethral glands lie dorsal to the pelvic urethra, caudal to the prostate and are embedded within the urethralis and bulboglandularis muscles.

## Vesicular glands

- paired, pyriform, thin-walled structures lying laterally to the ampullae, predominantly within the genital fold.
- sexual stimulation results in dilation and elongation of the vesicular glands:
  - up to 12–20 cm long and 5 cm in diameter.
- secretions make up the gel fraction of ejaculate.
- season influences the output of the vesicular glands:
  - gel fraction volume is highest during the physiological breeding season.
- individual stallions vary in amounts of gel fraction produced.
- ultrasonography, appear in longitudinal section as flattened oval/triangular sacs (Fig. 2.6).
  - shape depends on degree of sexual activity and time since last ejaculation.
  - amount and echogenicity of fluid within the glands is extremely variable:
    - both within and between stallions.
    - increased echogenicity is associated with the highly viscous gel fraction produced by some stallions.
  - thin echoic wall surrounds a uniformly anechoic lumen.

## Bulbourethral glands

- attach to dorsal surface of pelvic urethra, about 8 cm caudal to prostate gland.
- not usually palpable per rectum:
  - embedded in the urethralis and bulboglandularis muscles.
- secretions make up majority of pre-sperm or first fraction of the ejaculate:
  - likely function is to cleanse the urethra before ejaculation.
- ultrasonography, appear as oval structures with multiple small hypoechogenic spaces throughout the parenchyma (Fig. 2.7).
  - thin hyperechogenic line representing the gland wall is surrounded by the hypoechogenic bulboglandularis muscle.

## Ampullary glands

- enlarged distal portions of the deferent ducts.
- palpable along the midline of the pelvic floor over the neck of the bladder.
- ultrasonography, each ampulla is identified by:
  - hypoechogenic central lumen surrounded by uniformly echogenic wall and hyperechogenic outer muscular layer (Fig. 2.8).
  - uterus masculinus (remnant of müllerian duct system) is visible as one or two cystic structures located between the ampullae.

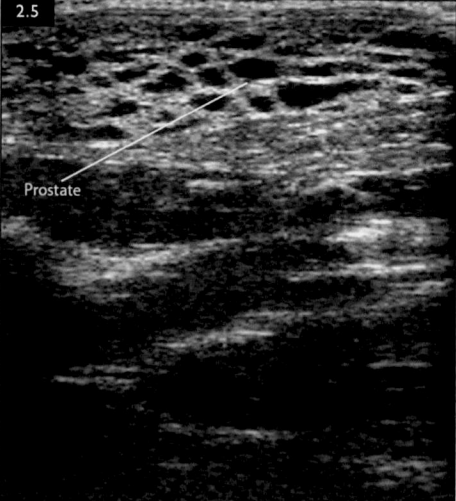

**FIG 2.5** Ultrasound image of the prostate of a stallion.

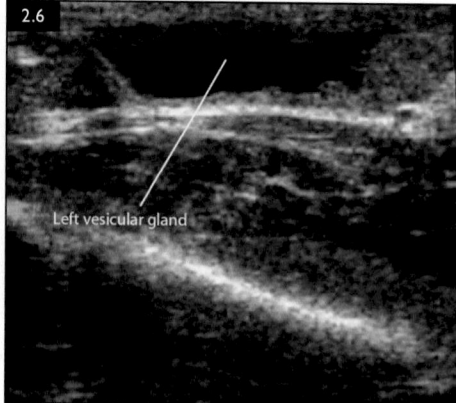

**FIG 2.6** Ultrasound image of the vesicular gland of the stallion, showing hypoechoic secretions accumulated within the lumen.

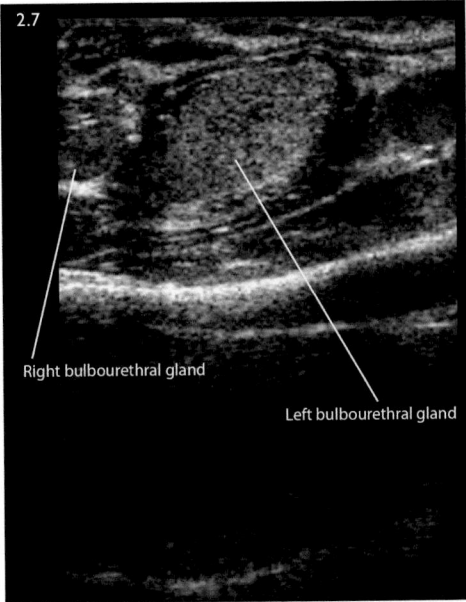

**FIG 2.7** Ultrasound image of the bulbourethral glands.

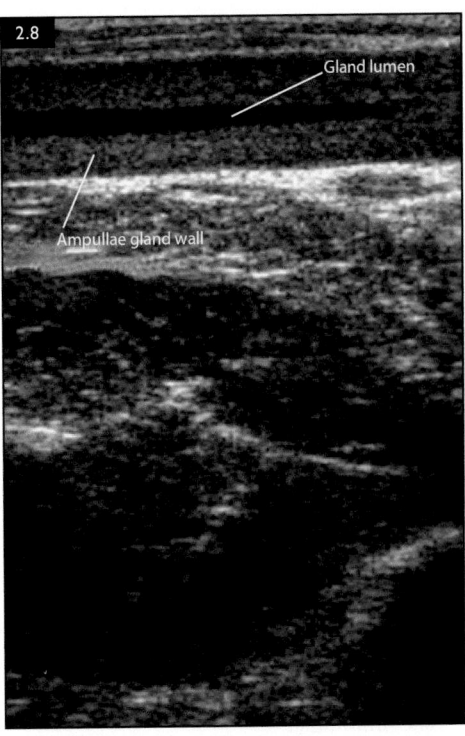

**FIG 2.8** Ultrasound image of the ampullary gland of the stallion.

## BREEDING SOUNDNESS EXAMINATION OF THE STALLION

### Handling a stallion for examination

- confident, experienced horse person should handle stallions for examination, breeding or semen collection:
  - aware of normal stallion breeding behaviour, including vocalisation, prancing, and arching of the neck:
    - ♦ do not attempt to discourage or correct these.
  - handler should work with the stallion away from the breeding area at first to become familiar with the stallion's behaviour and establish respect:
    - ♦ stallion should respond to:
      - – voice and lead corrections.
    - ♦ halt, back up, and turn when asked.
  - **excessive corrections, beatings, or discipline must be avoided:**
    - ♦ can establish long-term behavioural difficulties in stallions.
  - stallion must not be allowed to rush to the mare or phantom:
    - ♦ habit becomes difficult to break.
  - good fitting, preferably leather halter is essential:
    - ♦ long leather lead rope, preferably with a good quality chain and clasp, provides additional control.
    - ♦ chain can be placed over the nose and through the halter rings, or through the mouth if required.
    - ♦ once the handler is comfortable and the stallion under control:
      - – person conducting the examination can approach.

### Conducting the soundness examination

- palpation of external genitalia is best performed following semen collection:

- o  stallion more relaxed, willing to stand quietly, and more tolerant of examination.
- o  washing of penis prior to collection is approached in a similar manner.
- handler stands on the same side as the examiner:
  - o  clear communication between handler and examiner is essential.
- standing the stallion against one wall to limit his movement during the examination can be helpful.
- examiner should approach slowly but confidently from the left side at an angle to the shoulder, moving alongside the stallion until reaching the left flank.
- running the left hand or arm over the stallion's neck and back as one moves towards the stallion's flank, helps the stallion to be aware of the examiner's location.
- examiner should never surprise the stallion with movements or grasping of genitalia, as this may result in a kick with the hindlimbs.
- stallions can squeal and kick out behind when the scrotum or penis is palpated or washed:
  - o  with training, most stallions become accustomed to routine examination without difficulty.

# Introduction

- guidelines for evaluation of breeding soundness in stallions are published by the Society for Theriogenology (www.therio.org).
- primary purpose of the breeding soundness examination (BSE) is to select stallions:
  - o  expected to achieve pregnancy rates of at least 70% when bred to 40 mares by natural service or 120 mares by artificial insemination.
- BSE also used to diagnose causes of reduced fertility and develop management guidelines to optimise pregnancy rates.
- BSE includes:
  - o  accurate animal identification.
  - o  complete physical examination.
  - o  evaluation of libido.
  - o  bacterial culture of the urethra and semen.

- o  evaluation of semen for:
  - ◆  pH
  - ◆  total sperm numbers
  - ◆  sperm motility.
  - ◆  sperm morphology and longevity.
- o  examination of the external reproductive organs.
- additional tests that may be warranted during the BSE include:
  - o  examination of the internal reproductive organs.
  - o  endoscopy of the urethra.
  - o  measurement of motility after semen cooling or freezing.
- all results must be recorded clearly and permanently.
- following a BSE, stallions are classified as:
  - o  satisfactory, questionable, or unsatisfactory.
  - o  classification guidelines are listed in Table 2.1.
- stallions classified as questionable:
  - o  all or part of the BSE should be repeated after 60 days.
  - o  allows assessment of any changes or improvements in semen quality since the previous examination.
- **BSE results are not a guarantee of fertility/infertility and must not be represented as such.**
- stallions used in cooled transported or frozen semen programmes require additional testing not included in the standard BSE:
  - o  ensure adequate numbers of sperm remain viable after processing.

# Protocol for a breeding soundness examination

## Identification and history

- record the age, breed, and occupation of the stallion on a permanent veterinary record:
  - o  identifying the stallion using markings, tattoos, microchips, and/or photographs.
- detail historical information on general management including:
  - o  exercise, nutrition, hoof care, parasite control, disease, and lameness are noted.

**TABLE 2.1** Classification criteria for stallion breeding soundness evaluation (Society for Theriogenology)

**SATISFACTORY CLASSIFICATION**

- Good libido.
- Penis anatomically normal and free from inflammatory lesions.
- Bacterial culture does not result in pure growth of a single organism.
- Bacterial numbers decline from pre- to post-ejaculatory samples.
- Bacterial culture under correct conditions is negative for *Taylorella equigenitalus* (CEM).
- EIA (Coggins) test is negative.
- Two scrotal testes are present.
- Testicles and epididymides are of normal size, shape, and texture.
- Total scrotal width is >8 cm.
- At daily sperm output, the second of two ejaculates collected 1 hour apart contains a minimum of 1 billion progressively motile morphologically normal (PMMN) sperm.

**QUESTIONABLE CLASSIFICATION**

- At or below standard on two or more of the criteria above.

**UNSATISFACTORY CLASSIFICATION**

- Uni- or bilateral cryptorchidism.
- EIA (Coggins) test positive.
- Carrier of known genetic disease (e.g. combined immunodeficiency disease of Arabian horses).

- question the owner/trainer about any medications, supplements, or performance-modifying substances the stallion is receiving.
- obtain specific information on breeding history including:
  - number of mares bred per season.
  - overall seasonal conception rate.
  - average number of oestrous cycles mares are bred to achieve pregnancy.
  - non-return rates, and numbers of matings or inseminations per oestrus.
- obtain a description of the stallion's breeding routine and typical behaviour from the owner to provide insight into its management.

## Physical examination

- purpose is to ensure that the stallion has the desire and physical ability to breed.
- record the stallion's body condition and ability to ambulate normally.
- any condition that may impact on breeding ability should be noted:
  - lameness, debilitation, respiratory impairment, neurological disease, and partial blindness.
    - lameness caused by arthritis in the hindlimbs or spine will impair the stallion's ability or desire to mount and thrust during breeding.

- record any known heritable defects such as parrot mouth and aniridia.
- congenital unilateral or bilateral cryptorchid stallions are classified as:
  - unsatisfactory breeding prospects under the Society of Fertility and Theriogenology (USA) (SFT) guidelines due to the hereditary nature of the condition.

## Testing for infectious diseases

- collect serum samples for testing for venereally transmitted diseases such as:
  - Equine infectious anaemia (EIA) and Equine viral arteritis (EVA).
  - positive EVA serum antibodies:
    - submit semen samples for viral isolation or PCR (RT-iiPCR, or RT-PCR)
    - detect carrier stallions that are shedding virus in semen.
- collect bacterial culture swabs of the penile shaft, prepuce, urethra, and urethral fossa:
  - test for the presence of venereal pathogens:
    - *Pseudomonas aeruginosa*
    - *Klebsiella pneumoniae* (capsule types 1,2 and 5), and *Taylorella equigenitalis*.

## Evaluation of libido and mating ability

- where possible, perform the BSE with an oestrus mare present:
  - ○ properly assess the stallion's behaviour and libido.
- normal breeding stallion exposed to an oestrus mare should show:
  - ○ strong interest and achieve an erection within 1–5 minutes (Fig. 2.9).
- determination of adequate libido is subjective and based on many factors such as:
  - ○ season of the year, temperament, age, previous handling and breeding experience.

## Examination of the external genitalia

- penis and prepuce are most easily examined during teasing and washing for breeding:
  - ○ skin of penis and prepuce should be intact and free from erosions, crusts, or masses (Fig. 2.10).
- fossa glandis and urethral process should be examined:
  - ○ free from growths, ulcers or masses.
  - ○ fossa glandis checked for the accumulation of smegma, termed a 'bean'.
    - ◆ removed during washing.

- scrotal contents are most easily examined after semen collection when stallion is relaxed:
  - ○ scrotal skin should be free from erosions, crusts, and masses.
  - ○ testicles should be of similar size and consistency:
    - ◆ uniformly firm and freely moveable within the scrotum.
  - ○ head, body and tail of the epididymis should be palpated.
  - ○ if the tail and ligament are difficult to locate, testicular torsion should be suspected.
  - ○ spermatic cord is palpated running dorsally from testicle to the external inguinal ring.
- measure the scrotal contents:
  - ○ total scrotal width (TSW) is measured using calipers or ultrasound (Fig. 2.11).
    - ◆ larger TSW is associated with increased sperm output.
    - ◆ TSW correlates with testis parenchymal weight and daily sperm production.
      - – useful predictor of breeding potential.
    - ◆ average TSW for light horse stallions is 9–12 cm.

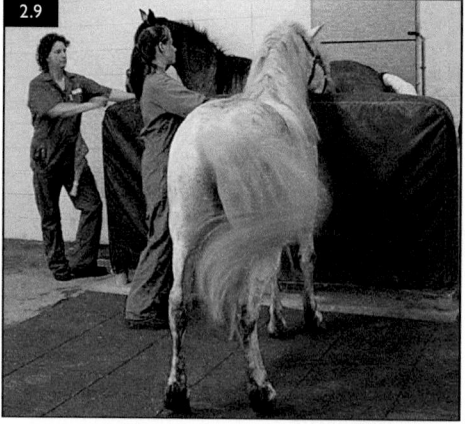

**FIG. 2.9** A maiden stallion being introduced to an oestrus mare over a padded board and encouraged to tease prior to training to the phantom for semen collection.

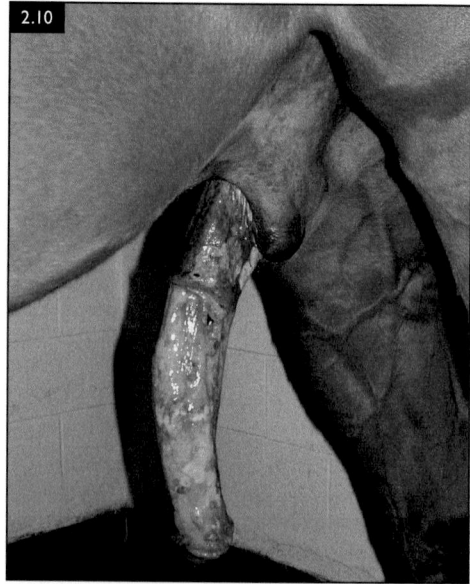

**FIG. 2.10** Lesions on the penis of a 16-year-old Thoroughbred stallion with squamous cell carcinoma. The disease had already spread to inguinal lymph nodes at the time of presentation.

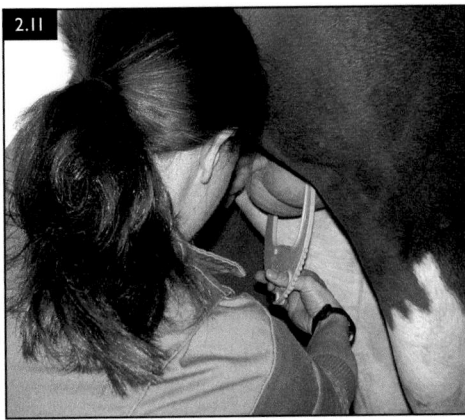

**FIG. 2.11** Measurement of the TSW of a stallion. The testes are pushed ventrally into the scrotum with one hand, while the calipers are placed over the widest part of the scrotum with the other hand.

- ♦ testis approximates the shape of an ellipsoid:
  - – calculate testicular volume: 4/3 π (length/2) (width/2) (height/2) or = 0.5236 × H × L × W in cm
- ○ testicular volume is then used to predict daily sperm production.
  - ♦ = [0.024 × (vol L + vol R)] - 0.76
- • predicted daily sperm production can be compared with actual daily sperm production to determine spermatogenic efficiency:
  - ○ where actual daily sperm production is below that predicted for testicular size requires further evaluation for disease conditions of the testes, epididymides, and accessory glands.

## Examination of the internal genitalia

- • examined by rectal palpation and ultrasonography:
  - ○ easiest after semen collection.
  - ○ aided by using stocks and sedation when necessary.
  - ○ phenothiazine tranquillisers should be avoided.
- • only the seminal vesicles and ampullae are palpable rectally.
- • all four accessory glands are visualised readily using ultrasonography.
  - ○ prostate gland located at neck of the bladder, about 10 cm cranial to the anal sphincter.
  - ○ vesicular glands are thin-walled, paired structures located craniolateral to prostate:
    - ♦ either side of ventral midline and may extend over the pelvic brim.
    - ♦ may be difficult to appreciate except during sexual arousal or following teasing (normal stallion) or if inflamed.
  - ○ paired ampullae are located cranial to the prostate and medial to the seminal vesicles:
    - ♦ easily palpable in the midline over the bladder neck.
    - ♦ approximately 1 cm wide by 12–25 cm long paired tubular structures.
    - ♦ not firm or distended (like a banana) unless there is ampullary blockage.

- ♦ stallions with TSW <8 cm may not produce enough sperm to meet BSE standard.
- ○ examiner stands close to the stallion's left flank and reaches under the abdomen to grasp both testicles within the scrotum with the left hand:
- ○ testicular size varies among stallions depending on:
  - ♦ breed, season, age, and reproductive status.
  - ♦ each testis of a post-pubertal stallion weighs between 150 and 300 g.
  - ♦ approximately 50–80 mm wide, 60–70 mm high, and 80–140 mm long.
- • ultrasonographic measurements may be more accurate than calipers.
  - ○ proper placement of probe essential to obtain a cross-sectional image and accurate measurements.
    - ♦ placed longitudinally across one testicle, while the examiner's left hand pushes the opposite testicle dorsally out of the way.
    - ♦ probe is turned to a cross-sectional axis to measure the width.
    - ♦ placing the probe across the pole of the testicle obtains the height measurement.

- internal inguinal rings can be palpated along the abdominal wall cranioventral to the pelvic brim and lateral to the midline:
  ○ bending and straightening the index finger while sliding the hand along the appropriate area of the abdominal wall will cause the finger to enter the ring.
  ○ ring can then be evaluated for hernia and adhesions.

## Bacterial culture

- after washing and drying the penis, the glans is stimulated to initiate the flow of pre-seminal fluid:
  ○ fluid allowed to wash the urethral lumen before urethra is swabbed with a bacterial culture swab.
  ○ second urethral swab is taken immediately after semen collection.
- an aliquot is removed from the semen, immediately post collection, for bacterial culture.
- normal stallions, bacteria cultured from samples are:
  ○ inconsistent and bacterial numbers decline after ejaculation.
- pure cultures or increased numbers of bacteria on the post-ejaculation sample are:
  ○ indicative of reproductive tract infection.

## Semen collection

- prior to a BSE, semen should be collected from the stallion daily for 7 days to deplete epididymal reserves and reach daily sperm output.
- for a BSE, semen is collected twice, with 1 hour between collections:
  ○ volume and pH of both collections should be similar:
    ♦ second collection contains approximately half the total sperm number.
  ○ any deviations from normal mounting, intromission, and ejaculation are recorded.
  ○ during semen collection. the strength of the ejaculatory pulses should be evaluated.
- semen must be collected and handled carefully to avoid:

  ○ cold shock or exposure to excessive heat (above body temperature).
  ○ direct light.
  ○ contact with spermatotoxins, such as lubricants, soaps and detergents.

## Hormonal evaluation

- hormonal evaluation within a standard BSE is the subject of some controversy:
  ○ normal stallions, where physical examination and semen analysis within normal limits:
    ♦ highly unlikely to provide any useful information.
- veterinarians managing stallions with large numbers of mares may elect to monitor hormonal status once or twice yearly.
- several samples are taken on consecutive days to avoid sampling in the trough of the normal pulsatile secretion.
  ○ serum samples can be tested for FSH, LH, oestrogens, testosterone, and inhibin.
- stallions with abnormalities on physical examination or spermiogram:
  ○ additional tests include hCG stimulation test or GnRH challenge test.
  ○ hormonal changes in stallions with testicular degeneration (TD) occur late in the course of the disease and may be most useful for prognosis:
    ♦ typical hormonal profile includes high FSH, low oestrogen and inhibin, and normal to low levels of LH and testosterone.
    ♦ oestrogen values below <124 pg/ml suggest irreversible damage to the seminiferous epithelium and a poor prognosis.

## Examination for venereal disease

- visual examination of the external genitalia:
  ○ penile, preputial and scrotal skin evaluated for erosions, granulomas, tumours, masses and exudates.
  ○ testis and epididymides are palpated to detect hard or soft areas and enlargements.

- bacterial culture of the genital skin, pre- and post-ejaculatory fluids, and semen:
  - after teasing, cultures of the prepuce, shaft of the penis and urethral fossa are collected.
  - erect penis is washed and dried thoroughly using warm water and soft cotton:
  - **washing the genitalia in disinfectant soap is generally contraindicated.**
    - avoids overgrowth with pathogenic or resistant bacteria.
  - after washing, the glans penis is massaged, and the pre-ejaculatory secretions are swabbed by insertion of a sterile swab into the urethra.
    - after ejaculation, the urethral swab is repeated.
    - in a normal stallion, the second swab demonstrates a reduction in bacterial populations compared with the first.
  - semen is collected for bacterial culture:
    - either from an open-ended vagina to allow fractionation.
    - or more usually from a regular artificial vagina (AV).

## Ultrasonography of the male genital tract

- scrotal contents and spermatic cords are easily evaluated using a portable ultrasound machine with a 5 or 7 MHz linear probe.
- testes are uniformly echogenic.
- epididymes are comparatively hypoechoic and granular.
- unilateral conditions can be evaluated by comparison with the contralateral side.
- biopsy of abnormal areas is facilitated by ultrasonography.
- ultrasonography of a cryptorchid testis and the ampullae and prostate glands is performed transrectally.
- cryptorchid testis can also be examined by parainguinal ultrasonography.

## Observation of libido and breeding behaviour

- libido and behaviour are observed during the stallion's typical breeding routine.
- sexually experienced stallions used for breeding in hand show strong interest in a mare in oestrus:
  - vocalising and prancing begin soon after the mare is visible.
  - most stallions will approach the shoulder of the mare and may sniff, then squeal and strike out.
  - if the mare is receptive, the stallion will move towards the mare's hindquarters, nuzzling the flank and perineal region before mounting.
- time required to obtain an erection depends on the stallion's previous breeding experience and temperament but is normally less than 5 minutes.
- stallions experienced in collection for AI, either by mounting a phantom or ground collection, also develop an erection rapidly during their usual breeding routine:
  - stimuli are stallion specific.
  - stallions should exhibit a strong drive to mount the phantom or service the AV.
- experienced pasture-breeding stallions may appear to show little interest in breeding:
  - when in an in-hand situation.
  - when the mare is not at the peak of oestrus.
  - test receptivity carefully by thorough teasing and mounting without an erection.
- after mounting, intromission of the penis into the vagina or AV stimulates strong pelvic thrusting.
- ejaculation occurs rapidly and consists of six to eight strong urethral pulsations.
- after ejaculation, penis almost immediately becomes flaccid, and the stallion relaxes.

## SEMEN COLLECTION AND EVALUATION

### Semen collection

- collection of semen from a stallion is best accomplished using a commercially available AV.
  - Missouri style, Colorado style, and the Hanover AV (Fig. 2.12).
- other methods of obtaining semen include:
  - collection of a dismount sample.
  - use of a condom or other collection device in the mare.
  - pharmacological ejaculation.
- AV is designed to hold hot water within a sealed jacket:
  - warmth/pressure of the AV on the stallion's penis causes ejaculation.
  - most stallions ejaculate readily with the temperature set in the range of 43–48°C (109.4–118.4°F).
  - temperature/pressure may be modified to suit a particular stallion's preferences.
- ensure that the stallion's penis is fully within the AV.
  - prevents ejaculated sperm from being exposed to the high temperature of the AV resulting in heat shock.
- AV is fitted with a collection bottle or bag:
  - fit an in-line filter to remove most of the gel fraction as well as dirt or smegma.

- immediately following collection:
  - place semen in an incubator maintained at 37°C (98.6°F) for initial evaluation.
- prepare and warm semen extender immediately before semen collection to 37°C (98.6°F) in the incubator.
- semen extenders provide:
  - energy source to support the high metabolic activity of spermatozoa.
  - antibiotics to combat bacterial contamination inevitable with collection.
  - protection from cold shock during cooling.
- semen from individual stallions varies in which type of extender it performs best:
  - test several extenders with every stallion to determine which suits it best.

### Semen evaluation

- raw stallion semen is very fragile:
  - poor handling can result in poor motility, morphology, and longevity.
  - rapid changes in temperature and exposure to water, detergents, lubricants, or other chemicals all impact negatively on sperm survival (Fig. 2.13):

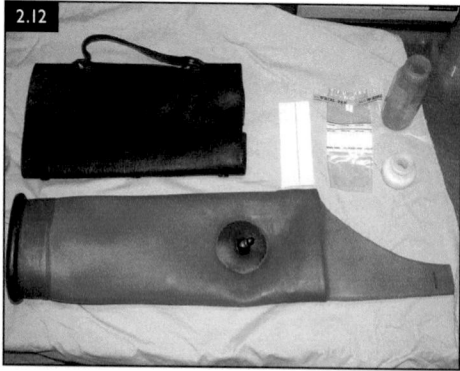

FIG. 2.12 The disassembled Missouri AV, including the rubber liner, leather case with handle, semen filter, collection bag, and bottle.

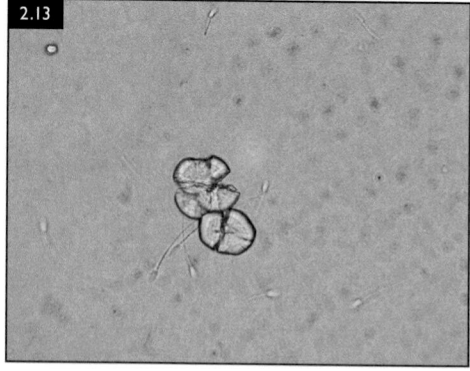

FIG. 2.13 These unusual contaminants found within a semen sample were determined to be starch granules and were the result of baby powder applied within the AV to prevent the rubber sticking together.

- all slides, coverslips, containers, pipettes, etc. contacting semen should be warm.
- after collection evaluate for:
  - colour and appearance.
  - volume and concentration.
  - total and progressive motility, and morphology.
- divide ejaculate into:
  - sperm-rich (produced by the testes and epididymides).
  - gel fractions (produced mainly by the accessory sex glands).
  - either use an in-line filter within the collection bottle.
  - or remove the gel fraction immediately after collection by filtering, decanting, or aspiration.
- **Semen colour**
  - white to skim-milk in colour and appearance.
  - colour approximates semen concentration:
    - watery represents a less concentrated sample than milky.
  - abnormal colour may indicate disease:
    - red semen indicates blood.
    - brown semen indicates blood or inflammation (pus).
    - yellow colour indicates urine contamination.
- **Semen volume**
  - record the volume of both the gel and sperm-rich fractions of the ejaculate:
    - measured using a warmed sterile plastic tube, clean dry baby bottle, or graduated cylinder.
  - normal semen volume varies with:
    - breed and age.
    - season due to changes in production of seminal fluid by accessory glands:
      - higher in the spring and summer, lower in the autumn and winter.
    - frequency of collection – decreases with more frequent collections.
    - time spent teasing the stallion prior to collection:
      - teasing increases the gel fraction.
  - typical average volume from an adult stallion is 50–70 ml (range of 25–300 ml).

- retain a small sample of unextended semen for:
  - morphology analysis and determination of sperm concentration.
- **multiply semen volume by sperm concentration to determine the total number of sperm in the ejaculate.**
- **Total and progressive motility analysis:**
  - good quality microscope with phase contrast optics and a heated stage improves the quality and repeatability of semen evaluations.
  - evaluate total and progressive motility on a raw semen sample, within 5 minutes of collection and prior to dilution with extender:
    - place a small drop of raw semen on a pre-warmed (37°C [98.6°F]) microscope slide with a coverslip.
    - evaluate several fields (up to 10) at ×200 to ×40 magnification with phase contrast.
    - average motility of the fields is recorded.
  - iSperm is a newly released semen analysis tool from Aidmics Biotechnology Co. Ltd:
    - allows an iPad Mini to be transformed into a handheld microscope with objective semen analysis software for the equine, available from the Apple Store.
    - more work is needed to improve the iSperm software for velocity measurements.
    - in present form, introduces a low-cost method for on-farm semen analysis (TM, PM, concentration) for breeders and veterinarians.
  - computer-assisted sperm motion analysis (CASA) systems are available:
    - they are expensive and less value in the field situation.
  - repeat the motility evaluation with semen extended in a 1:1 ratio, or to a concentration of 30–50 million sperm cells per ml.
    - permits evaluation of the motion characteristics of individual sperm.
  - **total motility**
    - **percentage of sperm cells moving, in any manner, in the sample.**
  - **progressive motility in the sample**

- ♦ percentage of sperm cells moving forward in a progressive manner.
  - ○ additional description of sperm velocity or vigour may also be given:
    - ♦ scale of 0 to 4.
    - ♦ 0 being non-motile and 4 describing fast or highly vigorous motility.
  - ○ retrograde or a tightly circular motion is abnormal.
- • pH
  - ○ evaluated within 10 minutes of collection using commercial pH paper:
    - ♦ normal pH is 7.2–7.8.
    - ♦ elevated pH may indicate urine or soap contamination or inflammation.
- • Sperm concentration.
  - ○ non-gel fraction can be determined using a haemocytometer, densimeter or Nucleocounter.
  - ○ samples contaminated with blood, urine, pus or debris will give spurious results.
  - ○ haemocytometer is a useful method for determination of sperm concentration:
    - ♦ dilute 20 µl of raw semen with 1.98 ml of formal saline or 10% buffered formalin or sodium citrate.
    - ♦ mix diluted sample (a 1:100 dilution) and draw up 10–20 µl into a pipette.
    - ♦ load sample by capillary action onto both sides of haemocytometer chamber:
      - − cover-glass already in place and be careful not to overfill it.
    - ♦ allow sample to settle for a short time:
      - − locate haemocytometer central 5×5 grid under the microscope at 20×.
    - ♦ count the number of sperm in all 25 squares within the central large grid.
    - ♦ repeat count on the other side of the haemocytometer:
      - − determine the average of the two counts.
      - − gives the number of sperm in 0.1 µl of the sample.
    - ♦ concentration of sperm in the original sample is therefore:

- − number of sperm in 0.1 µl × $10^6$/ml.
  - ○ total sperm number is multiplied by the percentage of progressively motile sperm:
    - ♦ gives the **total number of progressively motile** sperm in the ejaculate.
  - ○ during a BSE the sperm concentration in the second semen sample is expected to be approximately half that of the first sample:
    - ♦ concentration of sperm roughly the same as the first collection:
      - − consider whether complete ejaculation in both collections has occurred.
    - ♦ excessive drop in sperm concentration of the second collection:
      - − stallions with severe testicular or epididymal dysfunction or disease.
- • Sperm morphology
  - ○ determined either by examination of a fixed-stained specimen under oil immersion bright-field microscopy or by examination of a wet mount using differential interference-contrast microscopy (Figs. 2.14–2.18).
  - ○ most used stains are eosin-nigrosin or Hancock's stain.
    - ♦ others include modified Wright–Giemsa, Indian ink, Spermac, or new Methylene blue.
  - ○ drop of warm semen gently mixed with drop of warm stain at one end of a microscope slide:
    - ♦ smear is made using either a second slide or a glass pipette:
    - ♦ similar to the method used to make a slide for a differential blood smear.
  - ○ stained sample is allowed to dry.
  - ○ evaluate and count two hundred sperm under oil immersion microscopy:
    - ♦ record the number of normal/abnormal sperm within each category.
    - ♦ determine the percentage of normal and abnormal sperm in the ejaculate.
    - ♦ categorise the abnormal sperm morphology into type of defect present:

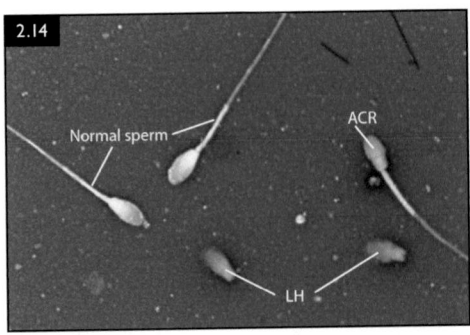

FIG. 2.14 Normal equine sperm. ACR = acromial head that has lost acrosome cap; LH = loose heads

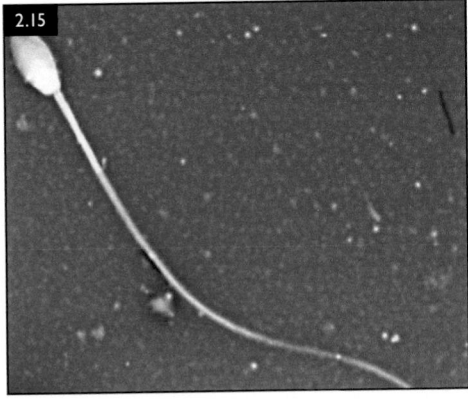

FIG. 2.15 Complete normal equine sperm with whole tail.

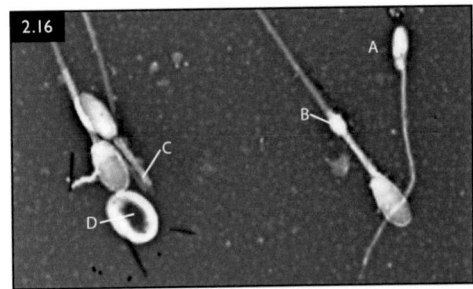

FIG 2.16 Abnormal equine sperm. A = small head with asymmetric tail position at attachment to head; B = distal droplet; C = bent midpiece; D = coiled tail.

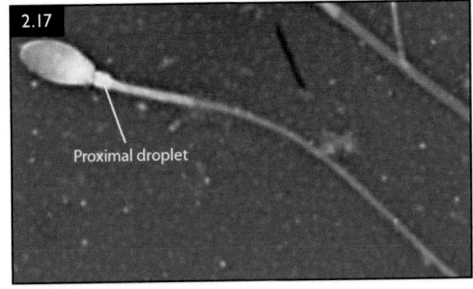

FIG 2.17 Abnormal equine sperm. Proximal droplet.

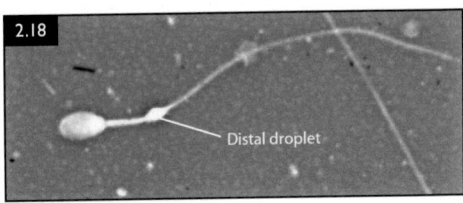

FIG 2.18 Abnormal equine sperm. Distal droplet.

    – head, midpiece, and tail defects.
    – proximal and distal droplets.
    – loose heads.
  ○ defects can be further classified:
    ♦ defects of spermatogenesis and
      testicular in origin (primary).
    ♦ occurring in the efferent duct
      system (secondary).
    ♦ caused by incorrect semen
      collection and/or handling
      technique (tertiary).
  ○ specific abnormalities may give insight
    into the presenting problem:
    ♦ detached heads in large numbers
      indicate:
      – stagnation of sperm in the
        ampullae, ductus deferens and
        epididymis.
    ♦ coiled tails are usually associated
      with significant testicular
      dysfunction.

    ♦ abaxial attachment of the midpiece
      to the head is common:
      – considered morphologically
        normal.
  ○ guidelines of the Society for
    Theriogenology:
    ♦ at least 60% morphologically
      normal sperm should be present
      in the ejaculate for a stallion to be
      considered a satisfactory breeder.
• **Sperm longevity**
  ○ **longevity of motility in raw semen:**
    ♦ held in a vial within in a 37°C
      (98.6°F) water bath.
    ♦ at least 10% progressive motility is
      anticipated after 6 hours.

- poor longevity associated with short sperm survival time in mare's reproductive tract.
  o assessed for **all stallions whose semen is being chilled and transported:**
    - extend semen sample to a concentration of 25–50 million/ml in several different extenders that vary in the sugar, buffers, and antibiotics used.
    - one sample from each extender is stored at room temperature and another stored in the transport container.
    - motility of semen in different extenders is examined at various times:
    - intervals might include 12, 24, 36, 48, and 72 hours, post collection.
    - results suggest whether a particular stallion's semen is likely to ship well over prolonged periods of time.
- **Number of progressively motile, morphologically normal sperm:**
  o calculation method is demonstrated in Table 2.2.
  o stallions classified as satisfactory:
    - second ejaculate contains more than one billion PMMN sperm.
- **Advanced tests:**
  o performed to evaluate other aspects of sperm function:
  o **Sperm chromatin structure assay (SCSA):**

- used to assess the denaturability of sperm chromatin.
- stallion with low fertility but all normal tests within the BSE are normal.
  o **Sperm membrane integrity:**
    - evaluated using the hypo-osmotic swelling test
    - fluorescent probes.
    - Nucleocounter SP-100 cytometry machine or with fluorescent stains.
  o **Acrosomal integrity** can also be assessed using specialised fluorescein staining.
  o **Scanning electron microscopy** can be used to investigate subtle structural abnormalities.
- **Chromosomal analysis**
  o normal chromosomal complement in the stallion is 64XY.
  o analysis may be required to investigate:
    - unexplained infertility in a young stallion
    - confirm chromosomal aberration in a suspected intersex condition.
  o aberrations of the autosomes are rare:
    - usually lead to embryonic failure early in development.
    - sex chromosome abnormalities are the most common finding.
  o two sodium heparinised tubes of freshly collected blood are submitted on ice overnight to cytogenetic laboratories for testing:
    - collected aseptically and without contamination of cells from other sources.
    - live lymphocytes are cultured for 72 hours.
    - nuclear chromatin collected and stained to perform the karyotype analysis.
    - specific C- and G-banding techniques used to identify specific chromosomes.
    - fluorescent *in situ* hybridisation and PCR techniques are used to probe for translocations, duplication, and deletions, including SRY expression.

| **TABLE 2.2** Number of progressively motile, morphologically normal sperm (PMMN) |
|---|

**SEMEN PARAMETERS**

| | |
|---|---|
| • Volume | 50 ml |
| • Concentration | 100 million/ml |
| • Progressive motility | 70% |
| • Morphology | 50% normal |

**EXAMPLE CALCULATION**

Volume × concentration × proportion motile × proportion morphologically normal

50 ml × 100 million/ml × 0.70 × 0.50 = 1, 750 million or 1.75 billion PMMN sperm

## PRESERVATION OF SEMEN

### Chilled semen

- transport of chilled extended semen is commonplace and greatly influenced the growth of AI.
    - when across state lines, countries, or continents may require that government permits, certification of disease-free status, or other paperwork is completed.
- commercially available transport packaging systems are now available, ranging from:
    - short term (24 hour) cardboard boxes with icepacks.
    - longer term (48–72 hour) Equitainer.
- post semen collection, semen volume, concentration and progressive motility are determined.
- semen is diluted with a pre-warmed extender to a concentration of 25–50 million sperm/ml.
- volume of extended semen required to provide a minimum of 500 million progressively motile sperm per dose is calculated.
- package the semen for chilling in a leak-proof container such as a Whirl-Pak bag, or all-plastic Air-Tite syringe.
- follow the directions of the shipping container regarding appropriate ice packs etc.
- clear labelling of the bag and the exterior of the container should include:
    - stallion name; owner; registration number; collection date.
    - motility; concentration; number of sperm shipped; mare identification.
- if semen will not reach the mare for 24–48 hours, factor in the expected decline in motility:
    - higher number of total sperm included in the shipment.
    - two doses of semen, each containing 500 million progressively motile sperm, are often sent.
- when receiving chilled-transported semen for AI there is no need to warm the semen prior to insemination.

### Semen freezing

- cryopreservation allows:
    - permanent storage of semen.
    - worldwide distribution of superior genetics.
    - continued breeding while a stallion pursues a performance career or is sidelined by injury or disease.
- cryopreserved semen appears to retain its fertilising ability virtually permanently when stored properly in liquid nitrogen.
- semen from individual stallions varies considerably in response to cryopreservation:
    - most stallions' frozen semen will result in acceptable fertility rates.
    - semen from some individuals cannot be frozen successfully:
        - reasons for individual variation are not known.
    - stallion with poor-quality semen is unlikely to be successful in a cryopreservation programme.
- per cycle pregnancy rates are generally lower than expected with natural cover, fresh, or transported breeding programmes:
    - average per cycle pregnancy rates with frozen equine semen are 50–70% range.
    - mares must be intensively monitored and inseminated very close to ovulation.
- cryopreservation process involves:
    - routine semen collection.
    - centrifugation to remove the seminal plasma.
    - re-suspension in freezing extender.
        - commercial, ready-to-use extenders greatly simplify semen preservation.
        - extenders contain buffers, antibiotics, sugars, egg yolk, and cryoprotectant.
        - protects the sperm from changes during freezing.
    - packaging, cooling and, finally, freezing in liquid nitrogen.
        - 0.5 ml plastic straw is currently the most widely used packaging.

- once frozen the straws are stored in liquid nitrogen until required.
- two types of containers are available for frozen semen:
  - dry-nitrogen shipper is a short-term storage and transport vessel:
    - liquid nitrogen is used to cool the vessel but is not kept within the it.
- models vary in the length of time that the temperature is maintained once charged (few days to 3 weeks).
  - liquid nitrogen tank is the best method for storage of frozen semen:
    - generally kept on the farm or in the veterinary clinic.

## ABNORMAL SEXUAL FUNCTION IN THE STALLION

- demands placed on many breeding stallions require them to function in an environment that is very different from the natural equine herd.
- the way stallions are housed and bred, plus the demands of performance, may cause extreme stress on certain individuals.
- prognosis for behavioural problems is good when there is a strong commitment from all the people interacting with the stallion.
- calm and consistent handling is the cornerstone of treatment.

## Self-mutilation

### Definition/overview

- act of a horse biting its own body, usually the flanks.
  - sometimes the chest or limbs.
- usually accompanied by squealing and kicking.
- most common in stallions, but it may also occur in geldings
- rare in mares.

### Aetiology/pathophysiology

- aetiology is unknown if all medical causes have been eliminated:
  - appears to be associated with frustration, particularly sexual frustration.
  - probably displacement behaviour.
  - stallion is motivated to court and breed mares or to fight other stallions.
    - if not possible, he bites himself.
- geldings pastured with their dams have an approach–avoidance response to their dam when she is in heat:
  - they may be motivated to breed with her if they are among the 25% of geldings who retain some sexual behaviour.
  - innate incest avoidance means they do not.
  - instead, they self-mutilate.

### Clinical presentation

- usual presentation is a stallion stabled with other stallions and with little opportunity to exercise.
- behaviour can lead to saliva-soaked hair to serious lacerations of the skin and underlying tissues.
- stallion can injure himself, and by kicking can also injure others.

### Differential diagnosis

- eliminate other causes:
  - painful conditions of the skin or underlying organs.
  - tapeworms or other parasites.
  - colic can present as biting at the sides:
    - usually, cow kicking directed at the belly.
  - neurological problems.

### Diagnosis

- biting at the body persists for weeks without worsening, unlikely to be colic.
- eliminate other causes of pain or pruritus.

### Management

- best treatment for stallions is provision of companionship, particularly a mare.
- isolate geldings.
- exercise, decrease in concentrates, and increase in roughage may be successful.

- cradle on the horse's neck if he bites his flanks (or a bib if he bites his chest) can prevent tissue damage:
  - horse may still swing his head to the flank in an attempt to bite.
  - can still kick, and may injure others.
- drug therapy has been successful in some cases:
  - opiate blockers, such as naloxone.
  - tricyclic antidepressant, amitriptyline, has been used in a gelding.

## Prognosis

- good if social change can be affected, but fair otherwise.

# Poor libido

## Definition/overview

- poor libido, or lack of sexual interest occurs when a stallion takes longer than expected to obtain an erection and ejaculate.
- normal stallions will promptly show strong libido upon exposure to an oestrus mare:
  - evidenced by vocalisation, prancing, interest in the mare, attainment of an erection and a strong desire to breed.

## Aetiology/pathophysiology

- lack of libido can be caused by a number of physical and psychological factors:
  - **novice stallions:**
    - sexual inexperience may result in slow sexual responses.
    - previous corrections and negative reinforcement for normal stallion behaviour displayed in undesirable locations.
    - endocrinological and physiological function is usually normal.
    - genetic selection for placid temperament.
    - stressful housing conditions.
    - previous punishment in the breeding shed.
  - **experienced stallions:**
    - negative sexual experience.
      - accidents such as kicks or falls.
      - chafing or burns caused by the AV or phantom.
    - overuse or use during the non-breeding season.
    - management changes.
    - orthopaedic pain or systemic disease.
  - TD in older stallions may result in low testosterone levels and low libido.

## Clinical presentation

- young stallions exhibit prolonged sniffing, nuzzling, and Flehmen behaviour:
  - slow to develop an erection and mount.
- inexperienced stallions may play and nip the mare without achieving erection:
  - lack focus and are easily distracted.
- experienced stallions may present with declining libido manifested as:
  - less vigorous teasing behaviour.
  - lack of focus.
  - more mounts per ejaculation.

## Differential diagnosis

- rule out erectile dysfunction and ejaculatory failure.
- thorough physical examination may determine an underlying issue causing pain.

## Diagnosis

- clinician must be familiar with the range of normal stallion behaviour.
- quiet observation of the breeding routine will often reveal the cause of poor libido:
  - preparation of the mare or breeding mount and AV.
  - stallion handling and washing.
  - teasing and mounting behaviour.
- experienced stallions:
  - detailed history focusing on recent changes in routine.
  - physical examination focusing on systemic, neurological or orthopaedic diseases.
- observation of the stallion's housing, turnout and exercise routines, and nutrition will help identify management issues contributing to poor libido.

## Management

- identification and correction of handling and management issues:

- usually resolve reproductive performance problems in young stallions.
- handler education and improved management are key to success:
  - providing light and airy housing.
  - encouraging exercise.
  - providing exposure to mares and eliminating exposure to other stallions.
- low libido stallions usually benefit from being housed in the mare barn:
  - testosterone levels, libido, and sperm production may increase significantly.
- teasing and breeding experiences should be positive and pain free:
  - patient handling techniques with minimal restraint.
  - safe and distraction-free environment.
  - exposure to mares in the peak of oestrus.
- stallions should be brought in hand to the breeding area and allowed to tease one or more oestrus mares for 30 minutes once or twice daily:

  - walking and circling the mare and allowing the stallion to tease face to face.
  - mount without an erection may improve arousal.
- hormonal treatment may be required in refractory cases:
  - GnRH (50 µg 1 and 2 hours before breeding)
  - increases testosterone levels and may increase libido.
- Table 2.3 gives indications and dosages for drugs commonly used in stallion reproduction.
- Table 2.4 gives the effects of some medications on reproduction in stallions.
- older stallions exhibiting declining libido:
  - appropriate management changes similar to those for young stallions.
  - orthopaedic disease should be treated appropriately with anti-inflammatories, shoeing, or joint injections.

**TABLE 2.3** Pharmacological agents used to alter sexual function in the stallion

| DRUG | DOSAGE | THERAPEUTIC EFFECT | ADVERSE EFFECTS |
|---|---|---|---|
| Testosterone propionate | 50–75 mg s/c every 2nd day Monitor T levels to keep below 2000 pg/ml Wean off slowly | Increase libido | Decrease semen quality; decrease endogenous testosterone production; increase aggression |
| GnRH | 50 µg s/c 2 hours and 1 hour before breeding | Increase libido by increasing endogenous testosterone | Frequent use or overdosage may decrease semen quality |
| Imipramine | 0.5–2.5 mg/kg p/o 2 hours before breeding | Lowers ejaculatory threshold | Mild sedation; dark coloured urine |
| Diazepam | 0.05 mg/kg (up to 20 mg max.) slow i/v 5 minutes before breeding | Reduces anxiety | Sedation and ataxia; disinhibition of aggressive behaviour |
| Phenylbutazone | 6 mg/kg i/v 1 hour before breeding or 2 mg/kg p/o q12h on an ongoing basis | Relieves pain | GI and renal damage with chronic administration |
| GnRH vaccine | Not documented | Temporary sterility; dosage, efficacy, and reversibility not evaluated in stallions | Does not eliminate sexual or aggressive behaviour in mature stallions; repeated dosing may result in vaccine reactions |

**TABLE 2.4** Performance-altering drugs with potentially negative effects on male fertility

| DRUG | COMMON USE | EFFICACY | ADVERSE EFFECTS |
|------|-----------|----------|-----------------|
| Progestagens (altrenogest) | Decrease sexual behaviour | Poor | Temporarily decreases testicular size, libido, testosterone levels and semen quality. May cause permanent decreased fertility if used in immature stallions |
| Phenothiazines (acepromazine, fluphenazine) | Tranquillisation, control unruly behaviour | Undocumented | Priapism, loss of erectile function |
| Reserpine | Control unruly behaviour | Undocumented | Penile paralysis, loss of erectile function |
| Anabolic steroids (boldenone undecylenate, nandrolone decanoate) | Improve athletic performance | Undocumented | Severe decline in testicular mass and semen quality |

- breeding process should be modified to provide maximum comfort:
  - ensure that the phantom or mare is at an ideal height.
  - training the stallion to collect from the ground.
- stallion's sexual workload should be minimised:
  - reduce the number of mares bred.
  - mares bred once at the optimum time during oestrus.
- low libido stallions may only be able to cover three mares per week:
  - compared with high libido stallions that may breed three or four times per day, 6 days of the week.

## Prognosis

- depends upon the cause of low libido.
- generally excellent in young stallions, with most improving significantly with patient training.
- good for improvement in mature stallions:
  - breeding schedule may need to be permanently reduced.

## Aggressive behaviour

### Definition/overview

- owner-reported aggression may in many cases be stallion exhibiting normal equine behaviour:

  - handling of the animal or the available facilities may be poor.
  - behaviour may escalate to unruliness because of inappropriate management.
- **stallions can be dangerous, and evaluation should be carried out by experienced personnel in a controlled environment.**

## Aetiology/pathophysiology

- aggression caused by a number of physical and psychological factors.
  - temperament is heritable:
    - some breeds and lineages are more challenging to handle than others.
  - psychological causes of aggression often based on frustration due to inappropriate management conditions:
    - unruly stallions may have minimal handling further increasing aggression.
    - may spend a large amount of time isolated from equine and human contact.
  - overuse of stallions and semen collection during the non-breeding season may result in increased aggression.

## Clinical presentation

- veterinary evaluation of an aggressive stallion often precipitated by an injury to a person or horse.

- stallions presented for aggression are boisterous/unruly in the breeding environment:
  - handling and management are often poor.
- common inappropriate behaviours include:
  - excessive biting and striking at the handler.
  - wheeling around and kicking out after dismounting.
  - charging towards the mare or phantom.
  - biting the mare during breeding.
- well-managed stallion with good manners can present with a history of unpredictable savage behaviour towards people or horses not associated with breeding:
  - rare and these individuals are dangerous.
  - may not be amenable to treatment.

## Diagnosis

- obtain a complete history of the stallion's management including:
  - type of housing, amount of exercise, and daily and breeding routines.
  - detailed description of the undesirable behaviour from the stallion's manager.
- observe the stallion's breeding routine and pay particular attention to:
  - restraint, handling techniques and corrections, and stallion's responses to them.
- conduct a complete physical examination, including:
  - palpation of the external genitalia, and lameness examination.

## Management

- retraining of aggressive stallions requires an experienced team in a safe facility:
  - handle the stallion with the minimum possible restraint equipment:
    - snug-fitting halter with a chain over the nose or a bridle with a snaffle bit.
    - dressage whip or 100 cm (39 inches) length of plastic pipe.
    - or a child's plastic baseball bat also works well.

- the whip, pipe or bat are not used for excessive punishment.
  - used to extend the arm length of the handler.
  - may provide direction while standing at the horse's shoulder.
- retraining starts in an environment that is not sexually stimulating and includes:
  - teaching the stallion to walk quietly in hand.
  - halt on command, back up, and move the haunches away from the handler.
- focus on positive reinforcement of appropriate behaviour.
- goal of the first collection or breeding:
  - improved behaviour and a calm and positive experience for the stallion:
  - use a docile, oestrus mare wearing appropriate protective equipment.
- decrease social isolation through introduction of a suitable companion horse:
  - daily grooming, turn out, and exercise decreases aggressive behaviour.

## Prognosis

- good with appropriate long-term retraining and consistent handling.
- stallions of limited breeding value:
  - castration may prevent human injury and improve the quality of life of the horse.
- properly managed stallions that exhibit unpredictable savage behaviour towards humans or other horses may warrant euthanasia.

# Erectile dysfunction

## Definition/overview

- inability of a stallion to obtain or maintain an erection, despite appropriate stimulation.
- stallions usually show good libido but do not attain a full erection.

## Aetiology/pathophysiology

- psychological factors.
- physical abnormalities of the genital or neurovascular systems:
  - previous injury during breeding, such as a kick from a mare:

- ♦ may result in permanent vascular damage leading to erectile dysfunction.
- ♦ penile paralysis, and paraphimosis following penile injury.
  - o priapism followed by penile fibrosis and paralysis following the administration of drugs such as phenothiazine tranquillisers or reserpine.
  - o occasionally, may appear to occur when the partially erect penis folds over and becomes trapped within the prepuce.

## Clinical presentation and Diagnosis

- history of traumatic injury and chronic paraphimosis.
- skin of the penis and prepuce is often thickened and tough, and sensation is impaired.
- fibrosis of the corpus cavernosum penis (CCP) results from:
  - o prolonged priapism, paraphimosis and ensuing thrombosis and oedema.
- during sexual arousal, the penis fails to become sufficiently turgid for intromission.
- most stallions have normal libido.
  - o those that have experienced repeated failed breeding attempts may show poor libido or frustrated breeding behaviour.

## Management

- where erectile function is permanently impaired but testicular function is normal:
  - o goal is to develop a protocol for reliable semen collection.
  - o provide additional stimulation during collection through manual pressure on the glans penis and application of hot compresses to the base of the penis.
    - ♦ lower ejaculatory threshold by administration of imipramine (2 mg/kg p/o) 2 hours prior to collection.
  - o chemical ejaculation may be attempted using:
    - ♦ imipramine (2 mg/kg p/o) followed 2 hours later with xylazine (0.66 mg/kg i/v).

- ♦ induces ejaculation within 20 minutes in approximately 50% of attempts.
- ♦ plastic bag is placed over the sheath.
- ♦ stallion is placed in a comfortable, quiet environment, such as his stall, prior to drug administration.
  - o with either method additional management can include:
    - ♦ exposure to oestrus mares in a distraction-free environment.
    - ♦ adequate time spent teasing.

## Prognosis

- stallions with chronic erectile failure due to physical or vascular abnormalities:
  - o poor prognosis for recovery and fair prognosis for fertility.

# Ejaculatory dysfunction

## Definition/overview

- any cause of abnormal sexual function resulting in a stallion that fails to ejaculate.

## Aetiology/pathophysiology

- most frequently caused by systemic abnormalities that hinder pelvic thrusting:
  - o arthritis of the spine and hindlimbs.
  - o neurological disease.
  - o aortoiliac thrombosis.
- abnormalities of the excurrent duct system can also prevent ejaculation:
  - o blocked ampullary glands.
- inappropriate temperature or pressure setting in the AV.

## Clinical presentation

- inability to obtain spermatozoa after collection with an AV.
- absence of urethral pulsing or tail flagging at breeding.
- sudden decline in pregnancy rate in stallions bred by natural cover.
- affected stallions usually show good libido and normal erectile function.
- stallions subjected to a cool AV:
  - o often continue to thrust, then stop and lose interest.
- stallions subjected to an AV that is too hot:

- often back out of the AV quickly, failing to ejaculate.
- cases caused by musculoskeletal or neurological disease may adopt:
  - atypical mounting positions.
  - tread and readjust the hind feet.
  - dismount after two or three thrusts and have a worried or painful expression.
- stallions with ampullary blockage show normal libido and pelvic thrusting:
  - urethral pulsation, tail flagging, and ejaculation do not occur.
- stallions occasionally present that appear to ejaculate normally:
  - few or no spermatozoa are present in emitted fluids.

## Differential diagnosis

- testicular or epididymal disease resulting in azoospermia.
- severely impaired spermatogenesis.
- retrograde ejaculation.
- congenital aplasia of the excurrent duct system.

## Diagnosis

- secondary to pain from orthopaedic disease, or due to neurological conditions:
  - complete physical examination to rule out these conditions.
- careful palpation of the external genitalia to detect:
  - excurrent duct abnormalities and testicular atrophy.
  - masses within the scrotal contents or spermatic cord.
- observe breeding behaviour and handling, including:
  - stallion's attitude, number of mounts, and position when mounted.
  - vigour of thrusting, and character of urethral pulsations.
- incorrect temperature of the AV:
  - correct range is 43–48°C (109.4–118.4°F).
- examine emitted seminal fluid for presence of:
  - spermatozoa or spermatazoal precursor cells.
  - alkaline phosphatase (AP) >1,000 IU/l indicate:
    - testicular and epididymal secretions are present, confirms ejaculation.

- obtain a urine sample by free flow or catheterisation:
  - retrograde ejaculation if high numbers of spermatozoa are found in the urine.
- palpation and ultrasonography of the urethra and accessory sex glands:
  - normal stallion the ampullae are 1–2 cm in diameter.
  - stallions with ampullary blockage larger and very firm on palpation.
- endoscopy of the urethra and colliculus seminalis may reveal abnormalities:
  - inflammation, purulent discharge, or physical obstruction.

## Management

- depends on the cause of the ejaculatory dysfunction.
- musculoskeletal and neurological disease are treated with appropriate medications and modification of the breeding routine to facilitate stallion comfort.
- ampullary blockage:
  - administration of oxytocin and transrectal ampullary massage prior to collection.
  - repeated collection attempts 2–3 times daily until ejaculation occurs.
- retrograde ejaculation may be treated with behaviour modification and imipramine.
- no causative abnormalities are detected:
  - treatment using a trial-and-error approach.
- unresolved cases:
  - testicular biopsy to determine spermatogenic activity.
  - cannulation of ductus deferens to determine patency can be performed surgically.

## Prognosis

- management issue identified and corrected, the prognosis is excellent:
  - incorrect phantom height, or improper AV temperature.
- ampullary blockage can be difficult to resolve:
  - after resolution, stallions return to previous levels of fertility.
  - sperm accumulation and blockage may reoccur after periods of sexual rest.

- advisable to collect affected stallions on a regular basis, even during non-breeding season.
- neurological or musculoskeletal disease impairing ejaculation:

    - function is improved by specific treatment.
    - prognosis for long-term fertility is guarded.
- aplasia of the excurrent duct system results in untreatable infertility.

## CONGENITAL ABNORMALITIES OF THE MALE REPRODUCTIVE TRACT

## Disorders of Sexual Development (Intersexuality)

### Definition/overview
- normal genotype of the horse is 64XX or 64XY.
- intersex is used to describe an animal with a sexually ambiguous phenotype.
- variations of intersexuality occur in the horse:
    - true hermaphrodites have both testicular and ovarian tissue present.
    - sex reversal describes an animal where the genetic sex disagrees with the gonadal sex (e.g. 64XX individual with testes).
    - pseudohermaphrodite describes an animal where the genetic and gonadal sex agree, but the presenting phenotype is in disagreement.
        - male pseudohermaphrodite:
            - 64XY with testes and female-appearing genitalia.
        - female pseudohermaphrodite:
            - 64XX with ovaries and male-appearing genitalia.

### Aetiology/pathophysiology
- number of disorders of sexual development result in phenotypic ambiguity:
    - monosomy X (Turner's Syndrome).
    - XXY Syndrome (Kleinfelter's).
    - mosaicism, gene translocations and receptor deletions.
- presence of SRY gene on the Y chromosome controls a cascade of events and testicular-secretory factors that control the development of the internal and external genitalia:

    - translocation of SRY gene from Y chromosome to an autosome may be responsible for intersex conditions in most cases.
    - absence of SRY gene in genetic males, results in failure of regression of the Müllerian ducts, which form the female tubular tract.
    - SRY-positive cases of XY sex reversal:
        - failure of androgen receptor expression on target organs and phenotypic female appearance.

### Clinical presentation
- variety of clinical presentations have been described:
    - ranging from infertility in otherwise normal appearing animals, to stallion-like behaviour with ambiguous external genitalia.
- often presented for veterinary evaluation due to:
    - inappropriate male-like behaviour in an apparent female.
    - abnormal appearing external genitalia.
    - high testosterone level following regulatory testing in a performance filly.
- Monosomy X, (64X0; Turner Syndrome).
    - mare intended for breeding may present for small stature and failure to cycle.
    - transrectal ultrasound examination:
        - small, inactive ovaries and small, flaccid, thin-walled uterus and cervix.
- Trisomy (65XXY; Kleinfelter's).
    - animals are male in appearance with either scrotal or cryptorchid testes.
    - sterile as ejaculates are azoospermic.
    - testes and penis are hypoplastic.

- True hermaphroditism.
  - individual with both ovaries and testes, either in the same or separate gonads.
  - rare in the horse (Figs. 2.19, 2.20).
  - exhibit stallion-like behaviour and have ambiguous external genitalia:
    - fused vulva, clitoromegaly and poorly developed or absent uterus.
    - retained abdominal ovotestes on palpation and ultrasonography per rectum.
- Female pseudohermaphroditism is very rare in the horse:
  - gonads are ovaries, and genotype is female (64XX).
  - external genitalia are male, often poorly developed and ambiguous.
- both XX and XY sex reversal have been reported in the horse.
  - XX individuals with testes are SRY-negative:
    - variable degrees of virilisation of female-appearing external genitalia.
  - XY individuals with ovaries have exhibited severe gonadal dysgenesis.
- Male pseudohermaphroditism.
  - most common disorder of sexual development in the horse.
  - individuals are 64XY:
    - abdominally retained testes and poorly developed female tubular tract.
  - wide variation in degree of virilisation of tract and phenotypic appearance:
    - generally female in appearance with stallion-like behaviour and abdominal testes.
    - clitoromegaly and increased anogenital distance.

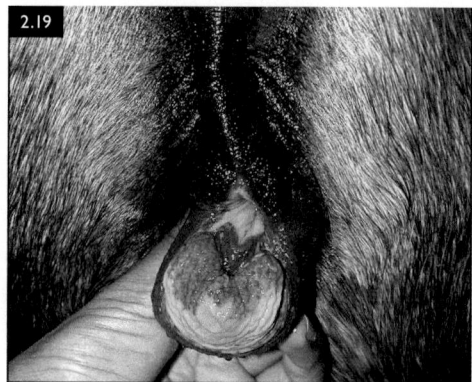

FIG. 2.19 View of the external genitalia of a true hermaphrodite. The animal presented as a filly with stallion-like behaviour and ambiguous genitalia. Note the presence of clitoromegaly. Chromosomal analysis revealed a 64XX karyotype. A blind-ending vagina was present, but no tubular reproductive structures were identifiable on transrectal palpation and ultrasound.

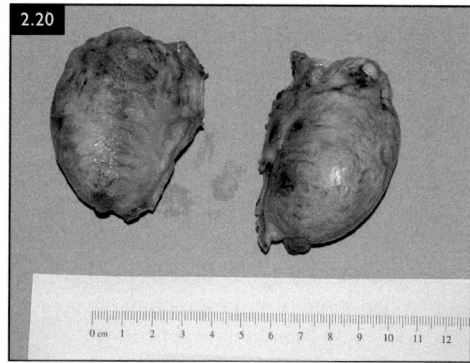

FIG. 2.20 Two abdominally retained gonads were present which were found histologically to be ovo-testes. The filly underwent gonadectomy and clitoridectomy and was retained as a pleasure mount.

## Diagnosis

- presumptive diagnosis based on clinical signs:
  - female-appearing animal exhibiting male behaviour.
  - infertile stallion with hypoplastic testes and penis.
  - ambiguous external and internal genitalia.
- definitive diagnosis requires karyotyping and additional cytogenetic analysis:
  - two sodium heparinised tubes of freshly collected blood.
  - submitted on ice overnight to cytogenetic laboratories for karyotype testing.
  - collected aseptically and without contamination of cells from other sources.
- differential diagnoses for stallion-like aggressive behaviour, in an apparent female, includes ovarian granulosa-theca cell tumour and exogenous steroid (testosterone) administration.

## Management and prognosis

- affected animals are usually sterile.
- surgical removal of retained testes and clitoridectomy should solve behavioural issues and allow the animal to be retained for performance or pleasure purposes.

## Cryptorchidism

### Definition/overview

- failure of normal descent of one or both testes into the scrotal sac.
- testis/testes located:
  - inguinally, abdominally, or subcutaneously outside the scrotum.
- testicular descent normally occurs between last 30 days of gestation and 6 months old.

### Aetiology/pathophysiology

- hormonal and physical events leading to testicular descent are not well understood.
- hereditary component is likely:
  - higher incidence of cryptorchidism occurs in the:
    - Standardbred, Quarter Horse, Percheron, American Saddle Horse.
    - Paint Horse, and Welsh Mountain pony.

### Clinical presentation

- unilateral cryptorchidism:
  - visual inspection and palpation of the scrotum reveal only one testis.
  - retained left testicle is more likely to be found abdominally.
  - retained right testicle is more likely to be found in an inguinal location.
- bilateral cryptorchidism:
  - no testes are palpable within the scrotum.
  - animal presents with unwanted stallion-like behaviour.
  - may have been sold as a gelding.
- apparent unilateral cryptorchidism is not unusual:
  - castration of the single scrotal testicle has been performed.
  - typically present as apparent geldings with stallion-like behaviour.

### Differential diagnosis

- previous castration of one or both testicles.
- true anorchid or monorchid condition is very rare.

### Diagnosis

- unilateral cryptorchidism:
  - inspection of the scrotum and careful palpation.
  - ultrasonography of the scrotum and inguinal canal.
  - testis may be retained within the inguinal canal or in the abdomen.
  - rectal palpation and transrectal/abdominal ultrasonography:
    - locating the retained testis in the abdomen (Fig. 2.21).
- bilateral cryptorchidism, or cases of unilateral castration of a scrotal testis with a retained abdominal testicle:
  - hormonal testing.
    - serum sample for baseline levels of testosterone and oestrogen.
    - testosterone values ≤0.24 ng/ml indicate absence of testicular tissue.

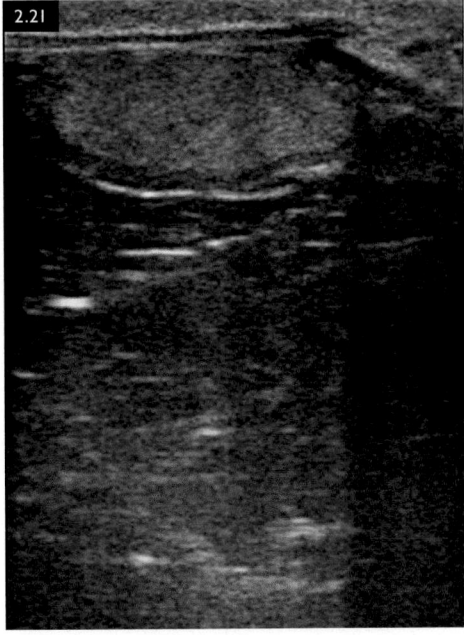

**FIG. 2.21** Transrectal ultrasound appearance of the retained abdominal testis in a cryptorchid horse.

- values ≥ 0.44 ng/ml indicate presence of testicular tissue.
- levels are low and stallion-like behaviour persists, or when values fall between 0.25 and 0.44 ng/ml, perform a hormonal stimulation test.
  - serum sample for baseline testosterone levels.
  - administer 5,000–10,000 IU i/v human chorionic gonadotropin.
  - follow-up serum samples are recommended:
    ◇ 60 minutes, 120 minutes, and 24 hours later.
  - significant rise (4–20 times) above baseline levels indicates the presence of testicular tissue.
  ○ total oestrogens assay may improve diagnostic accuracy in mature animals:
    - oestrogen or oestrone sulphate test alone is not recommended for colts less than 3 years of age, or donkeys of any age:
      - false negatives are common.
    - hCG stimulation test for both testosterone and oestrogen more reliable.
- development of an assay for anti-müllerian hormone (AMH) has proven useful for diagnosis of cryptorchidism:
  ○ concentrations were significantly higher in cryptorchids and intact stallions, compared to geldings (essentially undetectable).
- inter-laboratory variation means practitioners should interpret any test using the reference values supplied by their laboratory.

## Management

- surgical exploration and removal of the retained testis results in rapid improvement of unwanted stallion-like behaviour:
  ○ method chosen for surgical removal depends largely on the location of the testicle and the preference of the surgeon.
  ○ most common approaches include the inguinal, parainguinal, and laparoscopic (standing or recumbent).

## Prognosis

- exposure of the retained testicle to high temperatures results in increased likelihood of neoplastic transformation.
- affected stallions should not be used for breeding purposes due to the hereditary nature of the condition.

# Testicular hypoplasia

## Definition/overview

- underdevelopment of one or both testes.

## Aetiology/pathophysiology

- fairly common in the stallion and usually result of inherited genetic aberrations or chromosomal defects.
- may also result from cryptorchidism or exposure to teratogens, toxins, or possibly infections during foetal life.

## Clinical presentation

- unilateral or bilateral, and ranges from mild to severe.
- young stallion with testicular measurements well below the minimum recommended for its age may have testicular hypoplasia (Tables 2.5, 2.6).
- may have low libido, small testes, low sperm numbers, poor semen quality, and a history of infertility or subfertility.

**TABLE 2.5** Guidelines for total scrotal width by age for light horse stallions

|  | 2–3 YEARS | 4–6 YEARS | >7 YEARS |
|---|---|---|---|
| **Minimum** | 81 mm | 85 mm | 95 mm |

**TABLE 2.6** Guidelines for testis length and width for mature light horse stallions

| Dimension | Recommended minimum (SD) |
|---|---|
| Left width | 57.8 mm (5.2 mm) |
| Left length | 103.1 mm (82 mm) |
| Right width | 55.8 mm (5.8 mm) |
| Right length | 107.5 mm (8 mm) |
| SD = standard deviation | |

## Differential diagnosis

- acquired conditions of the testis such as TD.
- prepubertal stallions must not be erroneously diagnosed with hypoplastic testicles before growth is complete.
- **note it is common during reproductive evaluation of the stallion to observe one testis (usually the left) is larger than the other.**

## Diagnosis

- examination of a young, post-pubertal stallion with:
  - small testes, small epididymides, oligozoospermia or azoospermia.
  - history of poor libido and infertility.
  - testicular biopsy may be helpful in confirming the diagnosis.

## Management

- no effective treatment is known.

## Prognosis

- stallions with the condition appear to be predisposed to TD with advancing age.
- use of an affected stallion for breeding is discouraged since likely to be hereditary.

## CONDITIONS OF THE PENIS AND PREPUCE

## Phimosis

### Definition

- inability of the penis to be exteriorised or protrude from the prepuce.

### Aetiology/pathophysiology

- normal phimosis occurs in newborn colts:
  - penis is adhered to the internal prepuce for the first weeks of life.
- pathological phimosis usually occurs secondary to swelling and oedema following trauma:
  - kick from a mare during breeding (Fig. 2.22).
- also occurs with conditions that cause inflammation of the penis, scrotum, and prepuce:
  - coital exanthema, dourine, and neoplasia.
- significant ventral oedema from any cause may lead to :
  - severe oedema of the scrotum and prepuce:
    - result in prolapse of the internal prepuce.
    - trapping of the penis within the narrowed preputial ring.
  - once swollen, the effects of gravity worsen the oedema and the preputial prolapse, swelling, and oedema.

### Clinical presentation

- careful examination and palpation:
  - swelling is limited to the prepuce.
  - digital palpation usually reveals the tip of the penis within the prepuce.
  - chronic swelling, may lead to cellulitis, ulceration or sloughing of the skin.

### Differential diagnosis

- differentiate swelling and prolapse of the prepuce from involvement of the penis.

**FIG. 2.22** Severe preputial swelling and phimosis in a stallion following an accident while mounting a phantom with the AV affixed inside. The stallion had mounted the phantom aggressively and the prepuce became pinched between the edge of the AV and phantom, resulting in trauma.

- secondary phimosis may cause swelling which is unrelated to the penis, prepuce, or scrotum.

## Management

- oedema and swelling of the prepuce are treated by a combination of:
  - exercise
  - cold hydrotherapy (15–20 minutes every 4 hours).
  - local massage   ○ anti-inflammatories
  - diuretics.
- placement of a penile support sling:
  - reduces gravitational effects and helps relieve oedema (Fig. 2.23).

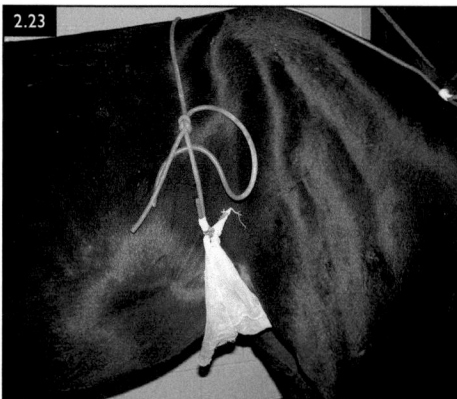

**FIG. 2.23** Sling applied to reduce effects of gravity on an oedematous scrotum and prepuce. The rubber tubing is placed over the back and additional tubing is run between the hindlimbs and tied over the back.

- apply Vaseline® or other emollient dressings to reduce damage to the preputial skin.
- diagnosis and treatment of the underlying condition is essential.
- preputial resection (reefing operation) is occasionally required to remove permanently diseased tissue.

## Prognosis

- good if caused by trauma:
  - condition improves quickly following initiation of therapy (Figs. 2.24, 2.25).

# Paraphimosis

## Definition/overview

- inability of the penis to retract into the prepuce.
- effects on penile circulation are rapid and severe.
- **treated as a veterinary emergency.**

## Aetiology/pathophysiology

- may be accompanied by severe preputial oedema following trauma:
  - rapid swelling of prepuce prevents the penis from retracting following detumescence.
- may occur secondary to penile paralysis, which may be associated with:
  - pelvic masses or neurological disease resulting in damage to the pudendal nerves and/or the retractor penis muscle.
- priapism, if not treated immediately, may also result in a secondary paraphimosis.

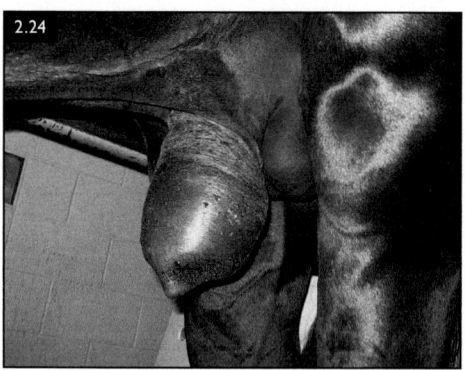 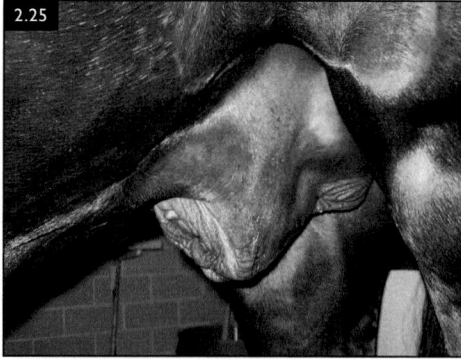

**FIGS. 2.24, 2.25** (2.24) Same stallion as in 2.22, 24 hours after initiation of therapy, which included hydrotherapy, emollients, a support sling, and anti-inflammatories. (2.25) Same stallion 48 hours after initiation of therapy showing resolution of the condition.

- once prolapsed, a rapid cycle of vascular compromise, oedema, and impaired lymphatic drainage ensues:
  - gravity effects on the pendulous penis further contribute to this cycle.
  - if left untreated, results in necrosis of the skin and gangrene of the penis.

## Clinical presentation and Diagnosis

- presents with a prolapsed penis that quickly becomes swollen.
- examine the penis for abrasions and preputial oedema if trauma is involved. (Fig. 2.26).
- may occur secondary to penile paralysis or priapism following administration of phenothiazine tranquillisers.
- with priapism:
  - penis is noted to be firm and partially erect.
- with partial or complete penile paralysis:
  - flaccid penis that cannot be retracted into the prepuce is observed.

## Differential diagnosis

- differentiated from simple preputial swelling and prolapse by careful examination of the affected area.
- evaluation of the history must include previous drug administration.

## Management

- **medical emergency in a stallion and immediate therapy is required.**
- acute condition:
  - gently replace the penis into the prepuce with the aid of sedation and restraint.
  - manual massage and application of a sugar or dextrose dressing assists with removal of tissue oedema, allowing replacement.
  - once returned to the prepuce:
    - apply retention device consisting of a taped, cut-off plastic bottle and rubber tubing.
    - penis is luricated with Vaseline® or other emollient dressing.
    - placed within the bottle with the urethral process aligned to the tapered bottle opening, to allow urination.

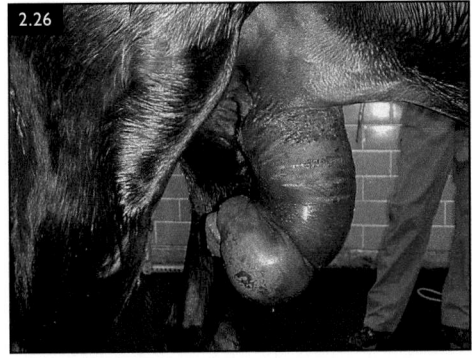

**FIG. 2.26** Penile haematoma and paraphimosis in a stallion following a kick from a mare during breeding.

- light padding is applied at the cut-off end of the bottle using tape or foam.
- open-ended bottle with the penis inside is then gently placed within the preputial cavity and held with the rubber tubing.
  - one set of tubing is passed over the stallion's back and tied.
  - other is passed caudally through the hindlimbs and tied dorsally over the back to the first set of tubing.
- stallion sling is then placed over the prepuce to ensure proper positioning.
- bottle and sling system should be removed twice daily for cleaning and to allow the penis to be inspected and re-lubricated.
- alternatively, the penis can be kept within the prepuce using a purse-string suture or clamps placed at the preputial ring.
  - may require sedation or general anaesthesia in the uncooperative animal.
- if penis cannot be immediately replaced:
  - use of a sling or similar supportive device limits the gravitational effects, which can worsen the oedema.
- combination of manual massage, hydrotherapy, exercise, and systemic anti-inflammatory therapy is indicated:
  - hydrotherapy should be limited to no more than 20 minutes of cold

application per session, repeated every 3–4 hours.

- ◆ longer application of cold (ice packs, cold hosing) can potentially cause more oedema as well as significant damage to the skin and tissues.
  - ○ gentle massage is not effective in alleviating penile oedema enough to allow replacement:
    - ◆ pressure wrapping with rubber bandages.
    - ◆ begin at distal end of the penis.
    - ◆ used temporarily to reduce oedema.
- **chronic cases.**
  - ○ penile skin becomes cracked, oozing, and eventually necrotic and gangrenous.
  - ○ chronic cases refractory to treatment, partial penile amputation may be required.

## Prognosis

- dependent on the rapidity with which veterinary treatment is sought and initiated.
- preputial and penile dysfunction can occur following successful treatment of prolonged cases due to fibrosis of the skin and damage to the vascular tissues of the penis.
- prolonged paraphimosis may also result in nerve damage leading to secondary penile paralysis.

# Penile paralysis

## Definition/overview

- inability to retract the flaccid penis into the prepuce.

## Aetiology/pathophysiology

- motor innervation of the retractor penis muscle via alpha-adrenergic fibres:
  - ○ most common cause of true penile paralysis in geldings and stallions is use of alpha-adrenergic blockers, such as the phenothiazine tranquillisers.
- also sequel to chronic paraphimosis, leading to nerve damage.
- other causes of reduced retractor penis muscle tone include spinal disease,

myelitis, tumours, severe malnutrition and exhaustion.

## Clinical presentation and Diagnosis

- complete penile paralysis:
  - ○ penis is flaccid and hangs from the preputial cavity.
  - ○ unable to retract the penis into the sheath despite being touched or exercised.
- manual examination reveals a flaccid penis that is often devoid of sensory sensation.
- lymphatic and venous drainage are disrupted with rapid progression of the condition.
  - ○ severe penile swelling, skin ulceration, and necrosis.
- fly strike may be a severe problem in the summer months.

## Differential diagnosis

- paraphimosis
- preputial oedema and trauma.

## Management

- initial treatment is aimed at reduction of swelling and oedema.
- placement of the prolapsed, paralysed penis in a sling apparatus:
  - ○ helps support penis and minimise gravitational effects on circulation.
- once replaced in the prepuce, can be held in place by a short-term purse-string suture surrounding the preputial opening.
- surgical treatment is indicated if treatment is unsuccessful:
  - ○ preputial resection (reefing operation).
  - ○ phallopexy (penile retraction).
  - ○ partial phallectomy (penile amputation).

## Prognosis

- return to full breeding capacity in stallions affected by penile paralysis is poor.
- chemical ejaculation using imipramine (2 mg/kg p/o) followed 1–2 hours later with xylazine (0.66 mg/kg i/v) may be possible:
  - ○ induces ejaculation within 20 minutes in approximately 50% of attempts.

# Priapism

## Definition/overview

- persistent erection, without sexual arousal.

## Aetiology/pathophysiology

- occurs due to continued filling/ engorgement of the corpus cavernosum with blood.
- phenothiazine tranquillisers are the most common cause of priapism in stallions and, less commonly, geldings:
  - alpha-adrenergic blocking properties block the sympathetic nerve pulses that initiate penile detumescence.
  - other tranquillisers and general anaesthetic agents less commonly associated with priapism.
  - once blood flow has been disrupted, a cycle of sludging of blood, disrupted outflow, oedema, thrombosis, and eventually fibrosis occurs.
  - chronicity, oedema and fibrosis further disrupt and occlude blood flow.
- exposure of colts to mares following phenothiazine tranquilliser administration may increase the risk of priapism.
- prolonged priapism:
  - secondary conditions such as paraphimosis and penile paralysis are likely.
  - due to damage to the pudendal nerves and retractor penis muscles.
- other causes of priapism include tumours, starvation, debilitation, spinal cord disease, and severe systemic illness.

## Clinical presentation and Diagnosis

- most commonly presents following phenothiazine administration with a firm, partially erect penis (Fig. 2.27).
- ultrasound imaging of the penis may demonstrate thrombosis.
- no history of phenothiazine tranquilliser administration:
  - comprehensive work-up including neurological examination to rule out spinal cord disease.

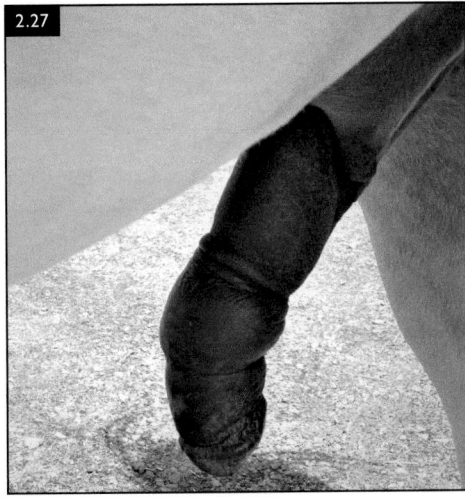

**FIG. 2.27** Priapism (sustained partial erection) in a pony gelding following repeated administration of phenothiazine tranquillisers.

## Differential diagnosis

- penile paralysis, paraphimosis and penile trauma.

## Management

- **must be treated as a medical emergency.**
  - early intervention is directed at limiting the dependent and gravitational effects.
    - massage, slings, and emollient dressings.
    - may prevent long-term sequelae if the erection subsides within a few hours.
- acute cases administer the ganglionic blocker, benztropine mesylate (8 mg i/v).
- administration of systemic anti-inflammatory agents may be helpful.
- chronic cases.
  - administer 10 mg of 1% phenylephrine HCl directly into the CCP under sedation or general anaesthetic.
- unresolved or more protracted cases:
  - surgical treatment under general anaesthesia may include:
    - lavage of the CCP with heparinised saline to remove sludged blood.
    - establish vascular shunts between corpus cavernosum and corpus spongiosum.

## Prognosis

- generally poor for stallions with priapism that has not resolved quickly.
  - damage to the pudendal nerve and retractor penis muscles may result in permanent penile paralysis, and fibrosis of the CCP over time.
  - results in erectile dysfunction.
  - penile amputation may be required in refractory cases.

## Penile lacerations/ trauma/haematoma

### Definition/overview

- trauma to the penis can cause a variable amount of injury:
  - minor swelling only.
  - significant haematoma of the vascular structures of the penis.
  - open lacerations with severe damage.
- **all penile injuries should be treated as immediate emergencies:**
  - rapid onset of tissue oedema and swelling.
  - potentially disastrous consequences for breeding future.

### Aetiology/pathophysiology

- commonly caused by a kick from a mare during attempted breeding:
  - penis is fully erect and at greatest risk of injury.
- tail hairs from the mare or suture material from a Caslick's stitch:
  - laceration of the stallion's penis during breeding.
- trauma during semen collection:
  - thermometer left in the AV.
  - stallion makes forceful thrusts against the phantom before the collector can properly place the AV.
  - phantom constructed with the AV inside predisposes to penile injury:
    - during mounting or collection, particularly if stallion slips and loses footing.
    - proper construction, maintenance, and use of the phantom are essential for the stallion's safety.
- significant potential for paraphimosis and/or penile paralysis exists.

## Clinical presentation and Diagnosis

- lacerations:
  - blood in the semen, dripping from the stallion's penis, or from mare's vulva after breeding.
- blunt force kicks:
  - swelling almost immediately and diagnosis is obvious (Fig. 2.28).
- thorough examination requires stallion's penis is exteriorised.
  - careful attention paid to urethral opening.
  - small lacerations are easily missed.

### Differential diagnosis

- differentiate from trauma to the prepuce.
- blood in the ejaculate – a urethral laceration a potential cause.

### Management

- penile injuries are emergencies and early treatment is essential for success.
- treatment options aimed at replacing the penis into the prepuce as soon as possible:
  - avoid secondary complications, such as paraphimosis.
- strict sexual rest until healing is complete.
- simple lacerations of the penile skin heal well without suturing if swelling and local infection are controlled.
- deep lacerations require general anaesthesia to permit surgical debridement and suturing.

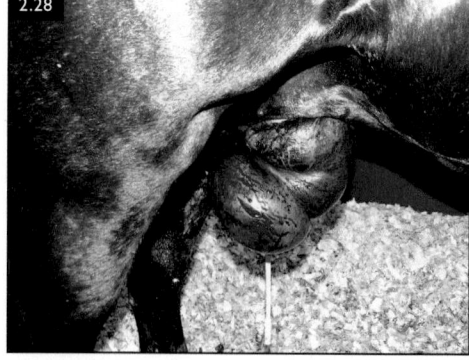

**FIG. 2.28** This Thoroughbred racehorse colt has sustained severe direct trauma to its prepuce and penis leading to extreme swelling and fresh bleeding lacerations of the penis and prepuce.

- o careful evaluation of the extent of the injury is necessary.
- o deep lacerations may involve the corpus cavernosum and the urethra.
- supportive care includes systemic broad-spectrum antibiotics and anti-inflammatory therapy.
- hydrotherapy is useful to reduce swelling and oedema.

## Prognosis

- simple injuries heal uneventfully with prompt treatment.
- more severe cases:
  - o recovery dependent on severity of the injury and involvement of corpus cavernosum or urethra.
  - o significant fibrosis can occur after haematoma formation and may result in:
    - ◆ penile deviations and incomplete erections.
    - ◆ ejaculatory dysfunction and penile paralysis.

## Smegma accumulation

### Definition/overview

- foul-smelling accumulation of secretions of sebaceous and sweat glands of the prepuce.
- forms on the penis and, particularly, within the fossa glandis.

### Aetiology/pathophysiology

- excessive accumulation of smegma may result in irritation and mild balanoposthitis.
- thick, wax-like material may form a 'bean' within the fossa glandis:
  - o result in discomfort and difficulty during urination or even ejaculation (Fig. 2.29).

### Clinical presentation

- foul-smelling odour from the stallion's or gelding's sheath.
- excessive straining during urination.

### Diagnosis

- observation of excessive accumulations of grey to black, thick substance on the penis and prepuce.

- examination of the distal penis and exploration of the fossa glandis reveals a 'bean' of smegma, which can be quite large.

## Differential diagnosis

- other causes of balanoposthitis:
  - o excessive force during washing of the penis.
  - o use of irritating soaps.
  - o infections such as coital exanthema.
- common causes of a foul smell from the prepuce:
  - o squamous cell carcinoma of the penis and/or prepuce.

## Management

- clean the penis thoroughly with warm water and cotton before each breeding/collection:
  - o periodically throughout the non-breeding season.
- copious amounts of water and limited manual pressure will reduce the irritation caused by washing.
- soaps and detergents are not recommended:
  - o disrupt the normal bacterial flora of the penile and preputial skin.
  - o predispose to growth of pathogenic organisms.
- accompanying balanoposthitis.
  - o removal of irritating smegma is generally an adequate treatment and the irritation resolves spontaneously.
- stallions used for regular breeding or semen collection:

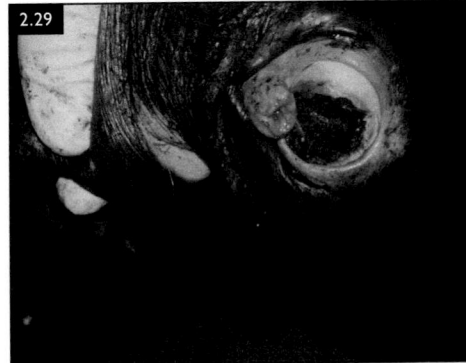

**FIG. 2.29** Smegma accumulation, also referred to as a 'bean', within the fossa glandis of a gelding that presented with difficulty in passing urine.

- excessive smegma accumulation is not a problem as routine washing of the penis.
  - beginning of the breeding season:
    - smegma accumulations may be significant.
    - require several washing cycles to remove.

## Prognosis

- good with correct general hygiene.

## Penile deviations

### Definition/overview

- exists when the erect penis persistently deflects inappropriately:
  - most often laterally or ventrally, making natural breeding difficult.

### Aetiology/pathophysiology

- uncommon in the stallion.
- usually result from fibrosis and adhesions within the vascular structures of the penis:
  - following traumatic injury such as penile haematoma.
  - often previous chronic paraphimosis.
- disruption of normal blood flow to one or more areas of the penis results in incomplete erection, while adhesion formation may result in deviation.

### Clinical presentation and Diagnosis

- history of penile injury.
- stallion may appear normal on physical examination during detumescence.
- observation of the stallion during teasing of an oestrus mare demonstrates a fully erect penis that deviates to one side or ventrally.

### Differential diagnosis

- incomplete erection due to inadequate sexual arousal.
- rupture of the suspensory apparatus of the penis.

### Management

- fairly recent injuries:
  - allow additional time for healing which may result in improvement.

- daily manual massage of the erect penis following exposure to an oestrus mare:
  - gently directing the penis to its correct position.
  - may help encourage blood flow to the affected areas.
  - efficacy of this approach is not clear.

## Prognosis

- guarded to poor depending on the chronicity of the condition.

## Rupture of the penile suspensory ligaments

### Definition/overview

- stallion's penis is attached to the ischial symphysis at its base by two short suspensory ligaments, which end continuously with the origin of the gracilis muscle.
- rupture of the penile suspensory ligaments is a rare condition.

### Aetiology/pathophysiology

- severe trauma during breeding results in rupture of the suspensory ligaments.

### Clinical presentation and Diagnosis

- history of severe trauma.
- significant ventral deviation of the erect penis is apparent.
- diagnosis is based on observation of ventral deviation and ruling out other causes.

### Differential diagnosis

- incomplete erection, and ventral deviation due to fibrosis following penile haematoma.

### Management

- recommendations for treatment are difficult due to the rarity of the condition.
- no reports of surgical correction exist.
- affected stallions may respond to pharmacological ejaculation to prolong reproductive usefulness.

### Prognosis

- guarded to poor.

## CONDITIONS OF THE TESTES, SCROTUM, AND SPERMATIC CORD

# Conditions causing scrotal enlargement

- variety of causes, including:
  - ○ tissue oedema     ○ trauma
  - ○ orchitis     ○ epididymitis.
  - ○ dermatitis     ○ tumours
  - ○ testicular degeneration     ○ torsion.
  - ○ varicocoele  ○ hydrocoele  ○ hernias.

## Scrotal oedema

### Definition/overview

- swelling of the scrotal wall.
- may be associated with:
  - ○ trauma (Fig. 2.30)     ○ infection
  - ○ haemorrhage
  - ○ other skin conditions, such as dermatitis or frostbite.
- typical finding in EVA infections.

### Aetiology/pathophysiology

- damage to the tissues of the scrotal skin or tunica leads to:
  - ○ extravasation and accumulation of fluid within the tissue.
- uncomplicated scrotal oedema seen in extremely hot weather – cause unknown.
- insulating effect of scrotal oedema impairs proper thermoregulation of the testis and can potentially impact negatively on spermatogenesis.

### Clinical presentation

- scrotum may be grossly enlarged, and the skin is thickened on palpation.
- semen quality is reduced due to effects on thermoregulation:
  - ○ reduction in sperm motility.
  - ○ increase in numbers of morphologically abnormal sperm.
- often seen with scrotal dermatitis.
- ventral oedema from other causes (e.g. hypoproteinemia) may have scrotal oedema.

### Differential diagnosis

- all other causes of scrotal enlargement including:
  - ○ trauma     ○ intrascrotal haemorrhage
  - ○ hydrocoele.
  - ○ scrotal hernia     ○ neoplasia
  - ○ testicular enlargement.
- scrotal and ventral oedema are classical findings in EVA and EIA infection.

### Diagnosis

- palpation of the enlarged scrotum reveals thickened skin and difficulty in differentiating the underlying testes and epididymides.
- ultrasonographic examination:
  - ○ increased thickness of echogenic skin layer, distinct from testis, in scrotal oedema.
  - ○ intrascrotal haemorrhage or hydrocoele:
    - ♦ accumulation of anechoic fluid between the scrotal skin and testis.

### Management

- treatment is dependent on the inciting cause.
- combat inflammation and swelling subsequent to trauma:
  - ○ systemic antibiotics, anti-inflammatories, diuretics, and cold hydrotherapy.
  - ○ care during hydrotherapy.
    - ♦ excessive cold application can be detrimental to blood flow and tissue/skin health, and impact on spermatogenesis.
    - ♦ limited to no more than 15–20-minute sessions every 2 hours.
- application of emollient dressings and a supportive sling in severe cases:
  - ○ may help reduce gravity-induced oedema in early acute stages of treatment.

### Prognosis

- good for complete recovery providing the inciting cause is treated appropriately.
- chronic scrotal oedema may disrupt normal spermatogenesis:

o altered thermoregulation of the testes may cause permanent effects on fertility.

## Scrotal and testicular trauma

### Definition/overview

- direct trauma to the scrotum can lead to lacerations, oedema, intrascrotal haemorrhage, and testis rupture.
- similar to penile injuries, scrotal injuries must be treated as immediate emergencies due to the potential for severe sequelae that may terminate a stallion's breeding career.

### Aetiology/pathophysiology

- most cases of scrotal and testicular trauma occur during breeding (Fig. 2.30, 2.31).

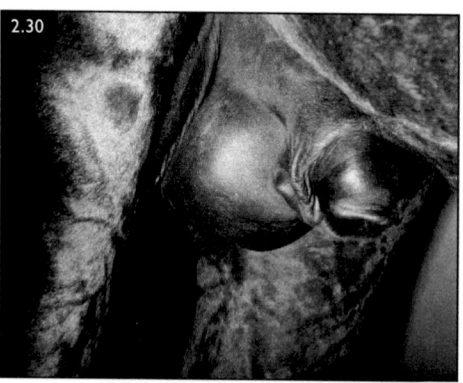

**FIG. 2.30** Swelling of the scrotal wall, which was associated with trauma from a kick by a mare.

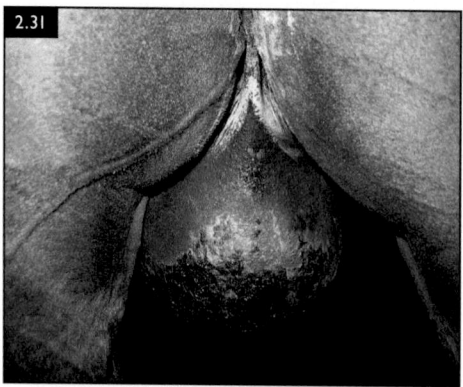

**FIG. 2.31** Stallion with severe scrotal cellulitis following a kick from a mare.

- other severe trauma such as a failed attempt to jump a fence.

### Clinical presentation and Diagnosis

- history may include a recent breeding incident.
- Paraphimosis and/or penile injury may also be present.
- scrotal lacerations may present with swelling and haemorrhage from the area.
- origin of the scrotal swelling is unknown.
  o diagnosis following thorough examination to rule out all other potential causes of scrotal swelling.
- ultrasound examination can be used to determine the severity of injury, the amount of blood accumulation, and whether testis rupture may have occurred (Fig. 2.32).

### Differential diagnosis

- consider all causes of scrotal swelling including scrotal oedema, scrotal hernia, orchitis, testicular tumour, testicular torsion, hydrocele, and peritonitis.

### Management

- scrotal trauma should be treated as an emergency.
- supportive therapy is as for scrotal oedema.

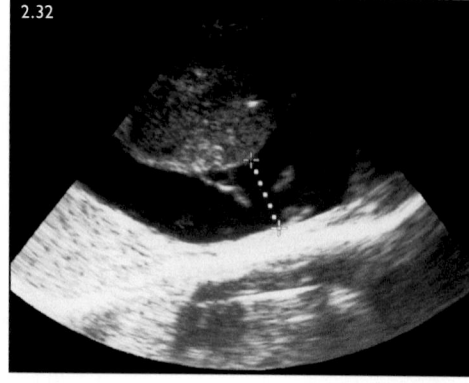

**FIG. 2.32** Ultrasound examination of the scrotum of a horse that had sustained a kick to this area. Note the presence of hypoechoic fluid (blood) between the parietal and visceral vaginal tunics (haematocoele). The testis has not ruptured.

- ultrasound examination suggests rupture of the testis from the tunica albuginea or significant blood clots within the scrotum:
  - surgical exploration is suggested:
    - tunic lacerations should be sutured separately.
    - fibrinous adhesions and blood clots can be removed.
- unilateral castration is recommended in cases of severe non-repairable trauma:
  - best option to save the contralateral testis.
- any injury accompanied by laceration:
  - surgical debridement and primary closure should be attempted.
  - severe swelling of the scrotum significantly complicates wound healing.
- broad-spectrum systemic antibiotics and systemic NSAIDs to control pain and inflammation.
- scrotal abscess with extension of existing scrotal infection to the peritoneal space can occur.

## Prognosis

- potential effects on spermatogenesis are significant:
  - increase in morphologically abnormal sperm occurring within days of injury.
  - azoospermia may occur about 2–4 weeks post injury.
- adhesions may develop between the testis and scrotum, resulting in permanent effects on thermoregulation and spermatogenesis.
- following appropriate therapy:
  - affected testis may slowly return to normal size and function 2–5 months following the injury.
  - stallions should undergo a complete BSE:
    - at least 60 days following complete resolution of the injury.
    - before the following breeding season begins.

## Orchitis/epididymitis

### Definition/overview

- inflammation of the testicle and/ or epididymis, causing testicular enlargement.

- often accompanied by fever and scrotal oedema.

## Aetiology/pathophysiology

- may be bacterial, viral, parasitic, or autoimmune in origin:
  - bacterial orchitis may be blood borne or by ascending/descending infection.
  - penetrating foreign bodies may cause local bacterial orchitis and/or abscessation.
  - viral causes of orchitis/epididymitis include EVA, EIA, and influenza.
  - migration of strongyle larvae has been associated with orchitis.
- primary bacterial epididymitis is rare.

## Clinical presentation and Diagnosis

- affected testicle(s) is (are) painful, hot, and swollen.
- orchitis is often accompanied by epididymitis and funiculitis.
- signs of systemic illness such as pyrexia, depression, and colic may be seen.
- thorough physical examination, including transrectal palpation to rule out inguinal hernia, is imperative.
- CBC reveals leukocytosis and hyperfibrinogenaemia.
- ultrasonography of the affected testicle reveals:
  - reduced echogenicity of testicular parenchyma and loss of normal echotexture.
  - scrotal oedema may be present.
- semen analysis demonstrates leukocytes within the semen.
- chronic epididymitis may be associated with abscessation and formation of adhesions.

## Differential diagnosis

- all other causes of scrotal and testicular enlargement considered including:
  - trauma, testicular torsion, haematocoele, neoplasia, and scrotal hernia.

## Management

- broad-spectrum systemic antibiotics chosen according to culture/sensitivity results.

- systemic anti-inflammatories and cold hydrotherapy to control inflammation and swelling.
- unilateral castration is recommended:
  - condition limited to one testicle.
  - not responsive to medical therapy.
  - increases the chance of retaining fertility in the remaining testicle.

## Prognosis

- future fertility in bilateral cases is poor.
- unilateral orchitis, the prognosis for the remaining testicle is guarded to good, provided therapy is initiated quickly.

# Testicular degeneration (TD)

## Definition/overview

- syndrome of progressive decline in fertility of the stallion accompanied by:
  - decreasing testicular size.
  - changes in testicular consistency towards softness.
  - significant reduction in semen quality.

## Aetiology/pathophysiology

- cause may be idiopathic or known.
- known causes include:
  - advanced age     ○ neoplasia
  - testicular trauma.
  - thermal injury (due to fever, swelling of scrotum etc.)
  - toxins     ○ radiation
  - administration of androgens or other hormones.
  - nutritional imbalance
  - vascular lesions
  - autoimmune diseases.
- some breeds or family lines within breeds appear predisposed to TD.
- horse appears predisposed to development of TD in middle age.
- regulation of spermatogenesis is a complex process involving the hypothalamic–pituitary–gonadal axis, Leydig cells, and the Sertoli cells:
  - any injury that causes disruption of these complex interactions may result in degeneration of the seminiferous epithelium of the tubules within the testicle and disturbances of spermatogenesis.

- degeneration may be focal or widespread, unilateral or bilateral, and stallions of any age may be affected.

## Clinical presentation

- initially, an increased testicular size and softness may be noted.
- with progression of the disease, the affected testicle(s) become(s) small and firm, with obvious wrinkling of the tunica albuginea.
- drop in fertility is noted.
- semen analysis demonstrates reduced sperm numbers, decreased motility and sperm longevity, and an increase in abnormal morphology.

## Differential diagnosis

- other causes of change in testicular texture should be ruled out:
  - including testicular hypoplasia and testicular neoplasm.

## Diagnosis

- history typically includes progressive decline in fertility and suspicion of poor sperm longevity.
- palpation of the scrotum reveals:
  - small, soft testes early in the course of the disease.
  - progression to a wrinkled tunic and small, firm testicles in advanced disease.
- ultrasonography of the testicles:
  - either normal
  - or demonstrates increased echogenicity of the testicular parenchyma:
    - fibrosis and calcification in the tubules.
- semen analysis demonstrates low numbers of motile, morphologically normal sperm.
  - advanced cases, total azoospermia may be present.
  - round cells (immature spermatids) are often found in the semen sample.
  - comparison of the stallion's expected daily sperm output based on testicular measurements, with actual daily sperm output following repeated collections:
    - reveals a disparity, consistent with poor spermatogenic efficiency.
- hormonal testing may assist with the diagnosis:

- baseline serum samples tested for FSH, LH, oestrogens, testosterone, and inhibin:
  - mean of several daily samples taken at the same time each day.
- typical hormonal profile of a stallion affected with TD includes:
  - high FSH.
  - low oestrogen and inhibin.
  - normal to low levels of LH and testosterone.
- oestrogen <124 pg/ml suggests irreversible damage to seminiferous epithelium.
- hCG stimulation test may be useful to determine the ability of the testicle to respond to LH stimulation:
  - two serum samples 60 and 30 minutes prior to injection of 10,000 IU of hCG.
  - follow-up samples are taken at 30-minute intervals for 3 hours.
  - stallions with TD typically have lower oestrogen and testosterone levels, following injection of hCG, than normal stallions.
- presumptive diagnosis can be made on the above findings:
  - definitive diagnosis requires testicular biopsy.

## Management and prognosis

- remove the inciting cause, if known.
- recent injury such as fever, trauma or administration of progesterone, can be identified:
  - allow 2–3 months for recovery prior to re-evaluation.
  - often demonstrate a significant improvement in the spermiogram.
- treatment of chronic idiopathic TD is controversial and has limited success:
  - GnRH injections via subcutaneous pumps – most widely employed therapy.
  - time and labour intensive and results are inconclusive.
  - owners should be forewarned that the condition is often irreversible.
- dietary supplements high in Omega 3 fatty acids may improve semen characteristics for some stallions.
- breeding management aimed at:
  - reducing the number of mares bred per season.
- careful monitoring of mares to ensure breeding close to ovulation is critical.
- continual monitoring of testicular size/character, and the spermiogram, will assist the stallion manager in optimising the fertility of a stallion with TD.
- additional management options for stallions with TD include:
  - processing of semen by gradient centrifugation to select the morphologically normal sperm for insemination.
  - use of low-dose deep horn insemination techniques.
  - may improve pregnancy rates enough to allow stallions with moderate to advanced disease to continue at stud for several additional seasons.
- most stallions with TD continue to progressively decline in fertility, and some eventually become azoospermic.

## Torsion of the spermatic cord

### Definition/overview

- spermatic cord rotates along the longitudinal axis:
  - often referred to as testicular torsion.

### Aetiology/pathophysiology

- orientation of the testis of the stallion in a horizontal fashion within the scrotum:
  - may predispose to more frequent spermatic cord torsions compared to other species.
  - factors allowing torsion to occur are not understood.
- spermatic cord includes the vas deferens, pampiniform plexus, and muscle and nerves.
- torsion of more than 180° rapidly leads to:
  - venous congestion, interference of arterial blood supply, and detrimental effects on the testis.

### Clinical presentation and Diagnosis

- 360° torsion of the spermatic cord is usually unilateral and presents with:
  - acute severe colic signs, scrotal enlargement, and reluctance to move.

- ultrasound findings of 360° torsion:
  - tail of the epididymis and caudal ligament are in their usual location.
  - caudal pole of the testis and the tail of the epididymis may be pulled dorsally due to the torsion of the cord.
- rectal palpation may reveal a thickened, painful spermatic cord within the vaginal ring.
- partial rotation of <180° is usually an incidental finding on reproductive evaluation:
  - usually transient and may recur in the same individual.
  - not accompanied by pain and no effect on semen quality.
  - readily diagnosed by palpation of the tail of the epididymis and its associated caudal ligament at the cranial pole of the testicle.

## Differential diagnosis

- main differential diagnosis is inguinal/scrotal hernia:
  - also presents with a stallion in acute pain with scrotal enlargement.
- other differential diagnoses include:
  - testicular neoplasm, epididymitis/orchitis, trauma, and haematocoele.

## Management

- rotation of 180° does not require any treatment.
- torsions of more than 180°, accompanied by clinical signs of pain and swelling:
  - **require immediate surgical intervention.**
  - affected testis usually cannot be saved:
    - ◆ unilateral castration recommended.
    - ◆ spermatic cord removed proximal to the origin of the torsion.

## Prognosis

- future fertility of the remaining testis is good provided:
  - condition is recognised and treated rapidly.
  - before thermal, ischaemia–reperfusion, or immunological injuries affect the other testis.

# Varicocoele

## Definition/overview

- abnormal distension and tortuosity of the veins of the pampiniform plexus within the spermatic cord.

## Aetiology/pathophysiology

- associated with infertility and reduction in semen quality:
  - postulated to be the result of inadequate cooling via the pampiniform plexus.
- varicocoeles are thought to arise from defects in the valves of the spermatic veins or from defects in the fascia surrounding the veins.

## Clinical presentation

- usually detected during routine BSE:
  - may be visible by inspection of the neck of the scrotum.
  - palpation of the spermatic cord reveals a lumpy texture.

## Differential diagnosis

- rule out that the spermatic cord changes are due to neoplasia of the spermatic cord, such as leiomyoma.
- other considerations include cord torsion and funiculitis.

## Diagnosis

- ultrasonography identifies enlarged vessels:
  - Doppler confirms the structures are vessels.

## Management

- no treatment is currently recommended if semen quality and fertility are good.
- unilateral orchidectomy is the only definitive treatment.

## Prognosis

- good if semen quality and fertility are good.

# Funiculitis of the spermatic cord (Fig. 2.33)

## Definition/overview

- inflammation of the spermatic cord, also known as scirrhous cord.

## Aetiology/pathophysiology

- usually a complication of castration.
- other causes in intact stallions include trauma, foreign body penetration, neoplasia, orchitis, and strongyle larvae migration.

## Clinical presentation and Diagnosis

- swelling and heat within of the spermatic cord.
- may be history of recent castration:
  - drainage of purulent material from the scrotum and pyrexia.
  - excessive granulation tissue with chronic infection can result in a very large cord stump (scirrhous cord).
- intact stallions with funiculitis may be an extension of orchitis/epididymitis:
  - present with a swollen, painful, firm cord and testicular enlargement.

## Differential diagnosis

- spermatic cord torsion
- spermatic cord neoplasia • varicocoele.

## Management

- post-castration funiculitis requires surgical excision with the incision left open to drain.
- postoperative broad-spectrum antibiotics and systemic anti-inflammatories.
- gentle walking exercise, and routine cleaning of the wound post-operatively.

## Prognosis

- good.
  - recovery may be prolonged in cases where there is considerable surgical dissection and tissue removal.

# Hydrocoele (Fig. 2.34)

## Definition/overview

- abnormal accumulation of fluid within the vaginal cavity, between the visceral and parietal layers of the vaginal tunic.
- with or without accompanying scrotal oedema.
- uni- or bilateral
- temporary or permanent in nature.

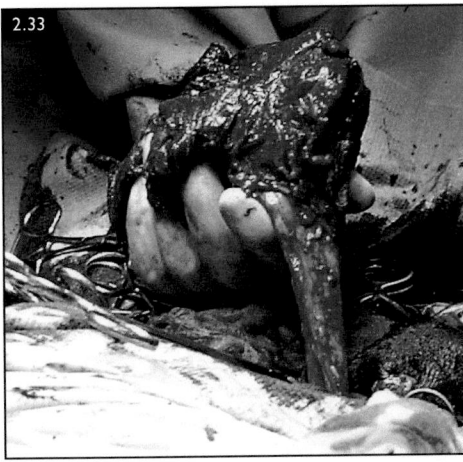

FIG. 2.33 This horse had undergone an open castration 4 weeks earlier and presented with a large swelling on the left side of the inguinal region. At surgery, the funiculitis lesion on the end of the vaginal cord is completely exteriorised. Note the normal cord below it.

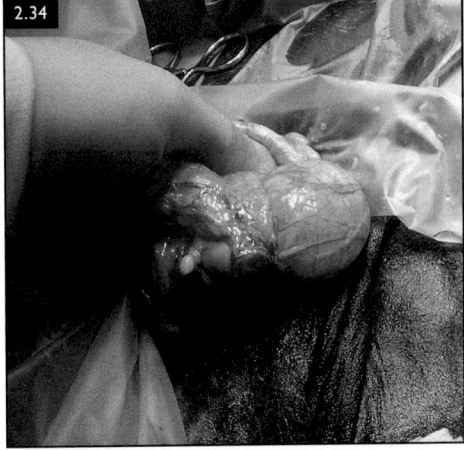

FIG. 2.34 Intraoperative view of a hydrocoele found at surgery following swelling in the scrotum of a horse castrated by the open technique several months earlier.

## Aetiology/pathophysiology

- vaginal tunic secretes a serous fluid to lubricate the vaginal cavity:
  - allows free movement of the testicle within the scrotum for optimal thermoregulation.
  - normally the fluid is reabsorbed by the lymphatic vessels of the spermatic cord.
  - any condition causing increased fluid secretion or decreased absorption will result in a hydrocoele.
- causes include inflammatory and non-inflammatory conditions:
  - trauma, neoplasia, orchitis, and extreme hot weather.
  - often idiopathic.
- vaginal cavity communicates with the peritoneal cavity:
  - peritonitis or ascites can cause hydrocoele.
- chronic or severe hydrocoele can impact on fertility:
  - insulating effect of the fluid and the subsequent effect on spermatogenesis.

## Clinical presentation and Diagnosis

- affected stallions present with unilateral or bilateral painless scrotal enlargement.
- systemic signs such a colic or pyrexia are absent.
- palpation:
  - testis and epididymis are freely mobile within the scrotum.
  - significant amount of fluid accumulation may obscure identification of structures.
- ultrasound examination demonstrates:
  - varying degrees of anechoic fluid within the scrotal sac, between testis and scrotal skin.
  - testis and epididymis appear normal ultrasonographically.

## Differential diagnosis

- all other causes of scrotal enlargement should be considered including:
  - cord torsion, trauma, haematocoele, scrotal hernia, and neoplasia.

## Management

- idiopathic hydrocoele:
  - drainage of the fluid is not recommended as not usually effective.
  - fluid reaccumulates quickly, and significant risk of bacterial contamination.
- cases occurring during periods of hot weather subside when the weather cools.
- moderate exercise may be helpful in alleviating fluid accumulation.
- anti-inflammatory therapy and diuretics of limited value.
- chronic hydrocoele can be treated by unilateral orchidectomy of the affected side.

## Prognosis

- guarded to good, depending on resolution of the inciting cause.
- most cases of idiopathic hydrocoele resolve spontaneously.

# Scrotal and inguinal hernia

## Definition/overview

- herniation of intestine and/or mesentery through the inguinal canal, via the inguinal ring, into the vaginal cavity.
  - herniation into the scrotum is termed scrotal hernia.
  - intestine that has penetrated the vaginal tunic, as may occur in the male foal during parturition, is termed a ruptured inguinal/scrotal hernia.

## Aetiology/pathophysiology

- inguinal hernias commonly occur congenitally in newborn foals, or soon after birth:
  - result of a large inguinal ring, allowing intestine to enter the vaginal sac.
- condition may be hereditary and can be unilateral or bilateral (Fig. 2.35).
- inguinal/scrotal hernias in the adult stallion are generally considered acquired:
  - affected stallions often have large inguinal rings palpable on rectal examination.
  - may predispose to the condition.
- Standardbred, Saddlebred, and Tennessee Walker horses appear to be at greatest risk.

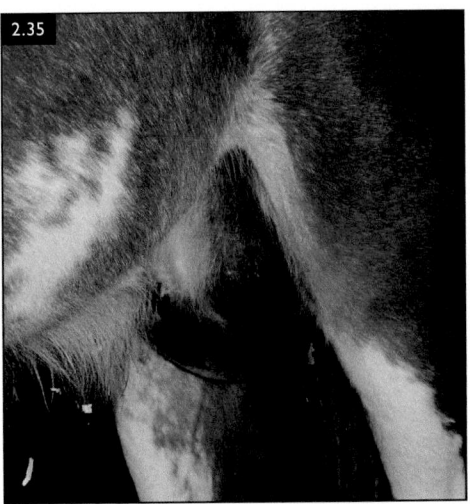

FIG. 2.35 Bilateral congenital inguinal/scrotal hernia in a young Clydesdale colt. Scrotal palpation revealed loops of bowel within the scrotal sac, which could be manipulated temporarily back into the abdomen.

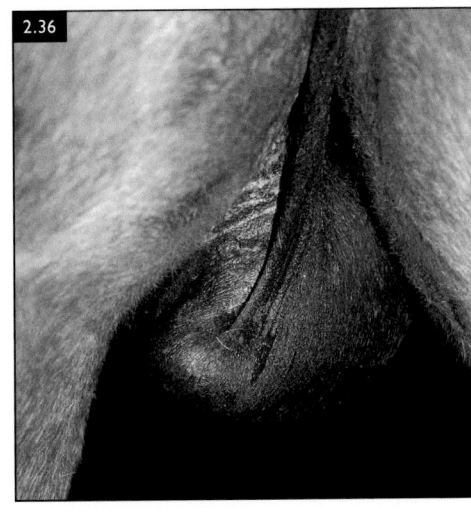

FIG. 2.36 A 3-year-old Arab colt showing colic and an enlarged, painful right testicle (caudal view).

- acquired herniation may be the result of increased abdominal pressure during falls, trauma, breeding, or exercise.

## Clinical presentation and Diagnosis

- foals with congenital scrotal/inguinal hernias present with:
  - unilateral or bilateral, soft, non-painful inguinal or scrotal enlargement.
  - at or soon after birth.
  - palpation of the inguinal region or scrotum reveals crepitation consistent with gas in the herniated intestine.
  - rarely, may become incarcerated and present with signs of pain.
- acquired scrotal/inguinal hernias in mature stallions usually present as:
  - severe, acute colic due to incarceration of the intestine.
  - often accompanied by scrotal enlargement (Fig. 2.36).
  - transrectal examination, distended bowel can be felt entering the inguinal ring.
- ultrasonography confirms the presence of bowel beneath the skin.

## Differential diagnosis

- all other causes of scrotal enlargement should be considered including:
  - torsion of the spermatic cord, orchitis, neoplasia and trauma.

## Management

- affected foals should be monitored for evidence of incarceration of herniated bowel:
  - some small hernias will self-correct before 3–6 months of age.
- manual reduction of the hernia with the foal in dorsal recumbency, and application of a support wrap (changed every few days) may result in resolution within 2 weeks.
- manual reduction is not possible:
  - ruptured hernia is likely.
  - hernia has become incarcerated.
  - surgical or laparoscopic reduction is required.
- **mature stallions with acquired inguinal/ scrotal hernias are emergency surgical colic cases.**
  - intestines quickly become severely compromised, often require resection and anastomosis (Figs. 2.37, 2.38).

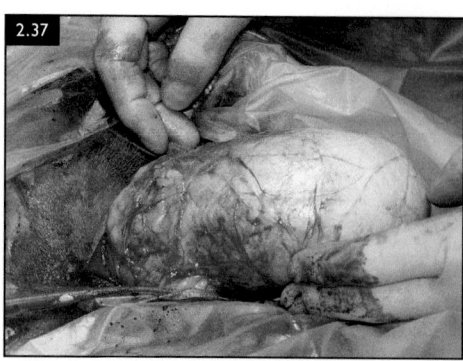

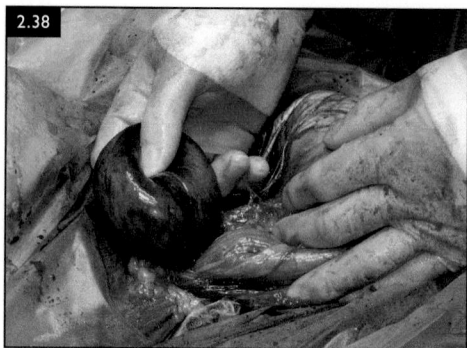

**FIGS. 2.37, 2.38** Same colt as in 2.36. Note the enlarged vaginal sac as it is removed from an inguinal incision (2.37). An incision through the parietal vaginal tunic revealed a small loop of jejunum that had been trapped and strangulated through the inguinal canal in the vaginal sac. This was manipulated back into the abdomen and resected via a separate midline laparotomy incision (2.38). The colt was castrated, and the superficial inguinal ring closed by interrupted sutures.

- ○ testis on the affected side of the hernia is often devitalised due to compression of the spermatic cord by the herniated intestine, requiring unilateral castration.
- ○ inguinal rings are sutured closed to prevent recurrence.

## Prognosis

- acquired inguinal/scrotal hernias carry a guarded to poor prognosis unless diagnosed and treated rapidly, due to rapid devitalisation of intestines and testis.

## Sperm granuloma

### Definition/overview

- granulomatous inflammatory reaction occurring secondary to trauma or infection:
  - ○ results in accumulation of cells into variably sized masses in the epididymis.
- relatively rare in stallions compared to other species.

### Aetiology/pathophysiology

- sperm that have escaped from the seminiferous tubules, excurrent ducts, or epididymis are highly antigenic, resulting in a surrounding granulomatous reaction.

## Clinical presentation and Diagnosis

- history of an affected stallion may include scrotal trauma, infection, laceration, or orchitis/epididymitis.
- palpation of the scrotum reveals single or multiple firm nodules of variable size in the region of the head, body, or tail of the epididymis.
- can cause complete obstruction of the epididymal lumen, resulting in azoospermia.
- diagnosis is based on history, clinical signs, ultrasonography (Fig. 2.39) and biopsy.

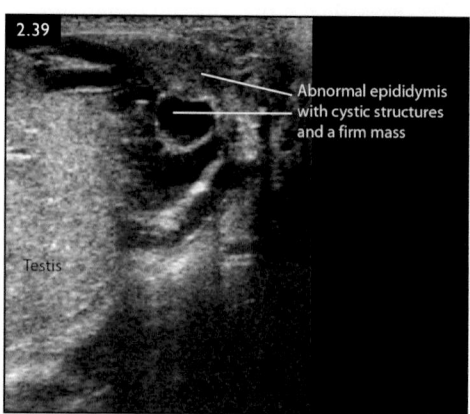

Abnormal epididymis with cystic structures and a firm mass

Testis

**FIG. 2.39** Ultrasound image of a sperm granuloma.

## Differential diagnosis

- neoplasia
- epididymal cysts.

## Management

- no known treatment to resolve sperm granulomas:
  - complete obstructions do not generally change with time.
- unilateral castration is indicated if sperm granuloma is confined to one testicle and azoospermia of the affected testicle is suspected.

## Prognosis

- poor for resumption of fertility if the condition is bilateral.

# Scrotal infection/cellulitis

## Definition/overview

- defined as a bacterial infection of the scrotal sac and/or skin.

## Aetiology/pathophysiology

- generally, follows scrotal trauma, but it can also occur secondary to peritonitis.
- chronic scrotal oedema can cause tissue damage, cellulitis, ulceration and eventual sloughing of scrotal skin.
- condition significantly impairs normal testis thermoregulation, affecting spermatogenesis.

## Clinical presentation and Diagnosis

- history may include trauma or exposure to a mare for breeding.
- present with an enlarged, swollen scrotum and pyrexia.
  - reluctance to breed due to pain.
- extension of peritonitis into the scrotal sac cases may present with depression and colic.
- diagnosis is obvious when infection is preceded by trauma with or without laceration.
- ultrasound examination of the scrotum will demonstrate:
  - scrotal oedema and pockets of fluid accumulation.

- abdominocentesis, a CBC, and fibrinogen estimation performed to determine the extent and severity of infection.

## Differential diagnosis

- all other causes of scrotal enlargement including:
  - scrotal oedema
  - scrotal sarcoid
  - scrotal hernia.
  - orchitis
  - neoplasia.
  - testicular torsion
  - hydrocoele.
  - frostbite of the scrotal skin.

## Management

- broad-spectrum systemic antibiotics and anti-inflammatories to control the infection and inflammation.
- chronic open wounds that do not involve the tunica albuginea should be:
  - debrided and lavaged with copious amounts of sterile saline under GA.
  - wounds are best left open to allow for drainage and secondary intention healing.

## Prognosis

- infection can be controlled and does not extend into the peritoneal cavity or the testis:
  - good.
- resumption of normal spermatogenesis may take 60–90 days.

# Scrotal dermatitis

## Definition/overview

- manifests as injury and inflammation of the scrotal skin.

## Aetiology/pathophysiology

- scrotal skin is delicate and prone to irritation from foreign substances such as:
  - leg paints, fly sprays, alcohol, and soaps.
  - many chemicals have the potential to cause contact dermatitis of the scrotal skin.
- overly aggressive cold application in cases of scrotal trauma can cause dermatitis.
- slight thickening of the scrotal skin due to dermatitis and its accompanying oedema

impacts on testis thermoregulation and spermatogenesis.
- chronic scrotal dermatitis can cause subfertility/infertility of stallions.

## Clinical presentation

- clinical appearance depends on the cause.
- slight thickening of the scrotal skin and oedema.
  - visible lesions of the scrotal skin.

## Differential diagnosis

- other skin conditions such as sarcoid, frostbite, ringworm, onchocerciasis, habronemiasis, or squamous cell carcinoma (SCC).

## Diagnosis

- potential reasons for dermatitis investigated by skin scrapings and skin biopsy, with the stallion under heavy sedation.

## Management

- removal of the inciting cause if possible:
  - Onchocerciasis can be treated by administration of ivermectin.
  - may result in transient worsening due to tissue reaction from dying microfilaria.
- Sarcoids may be surgically excised if they are not extensive.

## Prognosis

- guarded to poor, depending on cause.
- Sarcoids are difficult to treat, and recurrence is common.
- permanent effects on sperm quality are possible.

## POOR SPERM QUALITY

- poor sperm quality may consist of:
  - oligospermia (reduced sperm number in the ejaculate).
  - necrozoospermia (increased numbers of dead sperm).
  - teratozoospermia (increased numbers of morphological abnormalities of sperm).
  - azoospermia (absence of sperm in the ejaculate).
- may be caused by:
  - primary testicular disease, causing dysfunction of spermatogenesis.
  - extra-gonadal conditions of the epididymis, efferent ducts or accessory glands.
  - additional causes include environmental conditions, medications, illness, fever and poor semen handling.

## Spermiostasis/Ampullary Obstruction

### Definition/overview

- Spermiostasis is abnormal sperm accumulation during sexual rest within the excurrent duct system.
- accumulated sperm may partially or completely obstruct the ducts within the ampullary glands, efferent ducts and/or epididymides.
- 'Sperm accumulator' is a term sometimes used for stallions suffering this condition.
- unusually prolonged retention of sperm within the extra-gonadal reserves leads to oligospermia, necrozoospermia and/or teratozoospermia:
  - less frequently, complete azoospermia is seen.

### Aetiology/pathophysiology

- spermatogenesis continually produces new sperm in an assembly-line fashion.
- transit time through the epididymis takes approximately 7.5–11 days.
- in normal stallions, sperm are stored in the tail of the epididymis for only short periods of time and removed during ejaculation or masturbation.
  - senescent sperm also removed in the tail of the epididymis by phagocytosis.
- stallions prone to spermiostasis appear to have difficulty with the normal transit and emission of sperm from the epididymis:

- o allows sperm to accumulate there despite ejaculation.
- o physiological reasons behind this condition are not known.
- spermatozoa may occasionally accumulate in the distal ductus deferens and form obstructive plugs, leading to ampullary gland obstruction.

## Clinical presentation and Diagnosis

- affected stallions appear to ejaculate normally but have reduced sperm motility.
- ejaculate typical of spermiostasis contains:
  - o large number of tailless heads, distal droplets and hairpin bent tails.
- examination of the in-line filter following semen collection may demonstrate gritty casts of sperm clumps.
- spermiogram changes reflect the large number of degenerative, retained sperm in the ejaculate:
  - o may produce extremely large numbers of sperm on some occasions (e.g. 30 billion) and semen of unusually high sperm concentration (>500 million/ml).
  - o sperm and epithelial cell debris may form casts in the ejaculate.
- transrectal palpation and ultrasonography reveals enlarged and tense ampullary glands.

## Differential diagnosis

- other conditions causing alterations to sperm morphology such as TD or testicular neoplasia.

## Management

- stallions are best managed by increasing the frequency of breeding or collection:
  - o even if semen is not required.
  - o avoids spermiostasis during the breeding season.
  - o recurrence is common.
- several ejaculations over a relatively short period of time may be required to return semen quality to acceptable levels.
- frequent collection schedule should be resumed at least 2–3 weeks in advance of each breeding season to ensure that the semen quality is acceptable prior to the arrival of the first mares.

- transrectal massage of the ampullary glands and injection of oxytocin (20 IU i/m or i/v) or cloprostenol (125 µg i/m) immediately prior to semen collection may help alleviate a blockage.

## Prognosis

- good with appropriate management.

## Oligospermia/azoospermia

### Definition/overview

- oligospermia reduced numbers of sperm in the ejaculate.
- azoospermia is complete absence of sperm in the ejaculate:
  - o differentiate from failure to ejaculate.

### Aetiology/pathophysiology

- causes include frequent ejaculation, obstruction of the passage of sperm (e.g. ampullary obstruction or sperm granuloma), and decreased spermatogenesis due to disease (e.g. TD).

### Clinical presentation and Diagnosis

- oligospermia is characterised by repeated ejaculations with below normal levels of sperm, for the season of the year.
- height of the breeding season (May and June in the northern hemisphere):
  - o total number of sperm in the ejaculate should be >2.0–2.2 billion.
- non-breeding season (September to January in the northern hemisphere):
  - o total number of sperm should be >1 billion.
- careful observation during breeding or collection will confirm ejaculation has occurred:
  - o visualise tail flag.
  - o urethral pulsations felt with a hand on the ventral aspect of the base of the penis.
- submission of a sample of ejaculatory fluid for laboratory analysis of AP levels will differentiate between azoospermia and failed ejaculation:
  - o AP is produced in several locations within the male reproductive tract,

with the highest production in the tail of the eipididymis and testicle.

- ○ normal AP level in pre-ejaculatory fluid is 10–90 IU/l.
- ○ AP levels in ejaculatory fluid should be in the range of 1,600–50,000 IU/l.
- ○ stallions with apparent azoospermia and AP levels in the pre-ejaculatory range have failed to ejaculate.
- ○ stallion with apparent azoospermia and AP levels in the ejaculatory range can be considered truly azoospermic.
- ○ stallions with bilateral ampullary obstruction also have AP levels consistent with the pre-ejaculatory fluid range.

## Differential diagnosis

- incomplete or failed ejaculation.

## Management

- treatment depends on the cause of the oligospermia or azoospermia.
- azoospermia caused by complete obstruction due to epididymal sperm granuloma formation carries a poor prognosis.
- oligospermia caused by overuse can be resolved by:
  - ○ careful consideration of the number of mares that can be bred.
  - ○ compensatory reduction in the semen collection or breeding schedule.
- oligospermia caused by TD is best managed by reduction in the size of the stallion's book and optimal breeding management of mares.

## Prognosis

- guarded depending on the cause.

## Haemospermia

### Definition/overview

- presence of blood in the semen (Fig. 2.40).

### Aetiology/pathophysiology

- may arise from:
  - ○ wounds or lesions (habronemiasis or EHV) on the surface of the penis.
  - ○ ulcerations of the penile urethra or urethral process.

- ○ infections of the accessory glands, urethritis, urolithiasis, varicosities, neoplasia, and orchitis/epididymitis.

## Clinical presentation and Diagnosis

- thorough clinical examination as there are multiple causes.
- history may include:
  - ○ blood seen dripping from:
    - ◆ end of the penis.
    - ◆ mare's vulva following breeding.
  - ○ red or brownish discolouration of the semen following collection.
  - ○ urethral rent/laceration – frank blood may be seen following urination.
- presenting complaint may include reduced pregnancy rates due to reduced fertility.
- examination should include:
  - ○ inspection of erect penis and urethral process for small ulcers, wounds or SCC lesions.
- collection via an AV and cytological examination of the semen will confirm:
  - ○ presence of blood versus inflammatory cells, or both.
  - ○ seminal vesiculitis should be considered where haemorrhage accompanied by significant numbers of neutrophils,.
- most common cause of a significant amount of fresh, bright red blood in the ejaculate is a urethral rent/laceration:

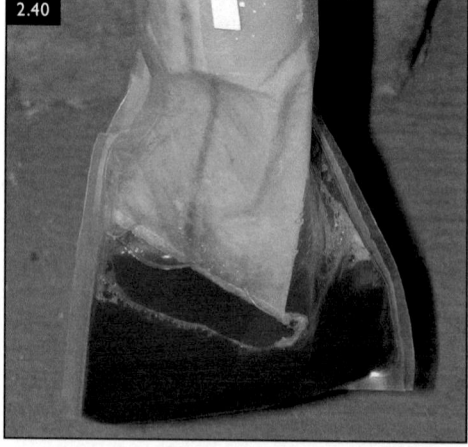

FIG. 2.40 Semen collected from a stallion with haemospermia. A urethral laceration was identified by urethroscopy.

o small urethral lesions can lead to significant haemospermia due to the communication between the lesion and the corpus spongiosum of the penis.

o urethroscopy using a flexible fibreoptic endoscope can identify urethral erosions or ulcerations, although this is not always easy.

o commonly found at the level of the ischial arch.

## Differential diagnosis

- blood observed following natural cover may originate from the mare, not the stallion:

  o mare should also be examined carefully.

## Management

- treatment depends on the source of the bleeding.
- strict sexual rest until non-neoplastic surface lesions of the penis are completely healed:

  o general wound care including keeping the wound clean.

  o application of topical emollient dressings may promote healing.

- deeper lacerations of the penis may require suturing under general anaesthetic.
- initial treatment of urethral rents involves strict sexual rest of about 8–12 weeks.
- urethral ulcerations that fail to heal despite prolonged sexual rest require:

  o temporary perineal sub-ischial urethrotomy to prevent continued haemorrhage through the urethral rent during urination.

  o systemic antibiotics and anti-inflammatories are useful adjunctive therapies.

- seminal vesiculitis is difficult to treat.

## Prognosis

- guarded depending on the source of bleeding and extent of the injury/infection.

## Urospermia

### Definition/overview

- presence of urine in the semen.

## Aetiology/pathophysiology

- smooth muscles of the bladder neck contract during ejaculation to prevent simultaneous urination.

  o any condition that affects the nerves and reflex arc responsible for ejaculation can result in urine contamination.

  o most cases of urospermia are idiopathic.

- potential diseases capable of causing this condition include:

  o hyperkalaemic periodic paralysis of Quarter Horses (HYPP).

  o cauda equina syndrome    o neoplasia.

  o EHV-1 infection.

- depending upon the degree of contamination by the urine:

  o rapid decrease in sperm motility and longevity – lowered fertility.

  o alterations in pH and osmolality of the seminal fluid.

## Clinical presentation and Diagnosis

- stallions may present with infertility or a sudden reduction in progressive sperm motility:

  o stallions used for natural cover – reduction in pregnancy rate is noted.

- semen collected for AI:

  o grossly contaminated with urine confirmed on examination for colour, odour, observation of urine crystals, and elevation of pH.

  o urospermia is often sporadic and contamination may not be seen with every semen collection.

  o creatinine levels >152.52 µmol/l (1.72 mg/dl) and urea nitrogen levels >10.71 mmol/l (30 mg/dl) in semen are highly suggestive of urospermia.

  o urine analysis (Azostix®) can confirm urine contamination of semen.

  o evaluation for the presence of inflammatory cells by staining a dried smear with a differential stain (Diff-Quik or similar).

- complete physical and neurological evaluation is suggested due to the potential for systemic illnesses to cause this condition.

## Differential diagnosis

- includes haemospermia and inflammatory products in the semen caused by orchitis, epididymitis or seminal vesiculitis.

## Management

- treatment of any underlying condition may reduce the frequency of urospermia.
- idiopathic urospermia:
  - train the stallion to urinate immediately before semen collection or breeding.
  - often accomplished by placing the stallion in a freshly bedded box stall a few minutes prior to semen collection.
- collecting semen into a bottle containing pre-warmed semen extender:
  - minimises the effect of urine on the sperm.

- ensure the extender is kept warm to prevent cold shocking of the sperm.
- imipramine hydrochloride (2 mg/kg p/o) given 2 hours prior to collection lowers the ejaculatory threshold, reducing the chance of urospermia.
- sympathomimetic agent phenylpropanolamine (0.35–0.5 mg/kg p/o q12h for minimum 14 days) before collection may increase bladder neck tone, reducing the chances of urospermia.

## Prognosis

- guarded since most cases are chronic and intermittent, responding only partially to therapy.
- where no cause is found:
  - stallions are best managed by encouraging urination prior to semen collection.

## CONDITIONS OF THE ACCESSORY GENITAL GLANDS

## Ampullary gland obstruction/ spermiostasis (see section on spermiostasis)

### Definition/overview

- obstruction is the most common condition of the accessory glands of the stallion.
- accumulation of sperm in the ampullae causes partial or complete:
  - unilateral or bilateral obstruction resulting in oligospermia or azoospermia.

### Aetiology/pathophysiology

- most often observed early in the breeding season or in stallions that are used infrequently for breeding.
- prolonged sexual rest can lead to the accumulation of masses of degenerating sperm and gel within the ampullary glands.

### Clinical presentation and Diagnosis

- complete, bilateral ampullary obstruction presents as azoospermia or, in the case of a stallion breeding naturally, as infertility.

- incomplete obstruction presents with semen of very low to zero motility, but high numbers of sperm.
  - large number of tailless heads, distal droplets or hairpin bent midpieces.
- diagnosis is based on the clinical presentation and a thorough reproductive evaluation:
  - rectal examination and ultrasonography demonstrate:
    - unilateral or bilateral enlargement and distension of the ampullae.
    - normal glands are soft on palpation and approximately 1–3 cm in diameter.
    - obstructed glands are firm, enlarged and distended on palpation (banana).
- measurement of AP activity in the ejaculated fluid:
  - differentiate ejaculatory failure, testicular-origin azoospermia and obstruction.
  - normal AP level in pre-ejaculatory fluid is 10–90 IU/l.
  - AP levels in ejaculatory fluid should be in the range of 1,600–50,000 IU/l.
  - stallions with apparent azoospermia:

- AP levels in the pre-ejaculatory range have failed to ejaculate.
- AP levels in the ejaculatory range has provided a complete ejaculation and can be considered truly azoospermic.
  - bilateral ampullary obstruction will have AP levels within the pre-ejaculatory fluid range.

## Differential diagnosis

- incomplete or failed ejaculation; azoospermia of other causes; TD as a cause of poor sperm morphology; poor semen handling as a cause of low sperm motility.

## Management

- low doses (10-20 IU) of oxytocin i/v within 10 minutes prior to semen collections.
- transrectal massage of the ampullary glands from a cranial-to-caudal direction prior to semen collection helps resolve the obstruction.
- frequent semen collections over a short period of time are usually required to resolve ampullary obstruction.

## Prognosis

- excellent with appropriate management:
  - usually reversible and stallions return to previous levels of fertility.
- may recur in some stallions:
  - maintain regular, frequent collections throughout the year to prevent recurrence.

## Accessory genital gland infections

### Definition/overview

- most common accessory genital gland infection encountered is seminal vesiculitis:
  - rare cases of prostate infection/abscess have been reported.

### Aetiology/pathophysiology

- most common causative agents include *Pseudomonas aeruginosa, Klebsiella*

*pneumoniae, Streptococcus* spp., and *Staphylococcus* spp.
- infection may occur due to blood-borne infection, ascending infection, extension of orchitis/epididymitis, or secondary to urethritis or cystitis.

## Clinical presentation and Diagnosis

- stallions may present with a history of declining fertility, a reduction in semen quality, or ejaculatory dysfunction.
- acute infections:
  - may present with painful ejaculation or reluctance to breed, depression, and inappetence:
  - other systemic signs are uncommon, and pyrexia is not a typical finding.
- collect semen into an open-ended AV to allow fractionation of portions of the ejaculate:
  - different fractions may identify the source.
  - infection may/may not be accompanied by gross changes of colour in the ejaculate.
  - examination of the ejaculate with a differential stain demonstrates polymorphonuclear (PMN) cells and bacteria in the semen. (Fig. 2.41).

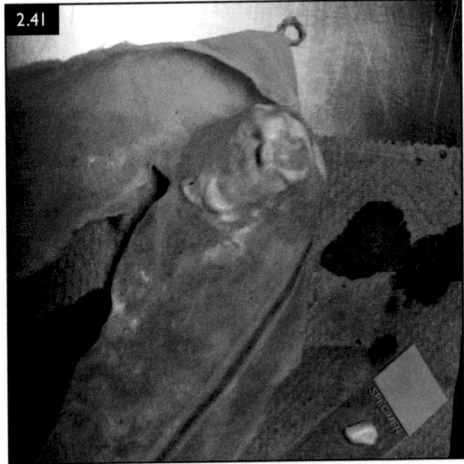

**FIG. 2.41** Semen filter after semen collection from a 10-year-old Arabian stallion suffering from ampullary obstruction. The stallion was azoospermic until this gritty mass of inspissated sperm was ejaculated. The stallion's spermiogram continued to improve with each follow-up semen collection.

- o aerobic culture of the semen demonstrates pure growth of pathological bacteria.
- thorough reproductive evaluation including:
  - o examination of the penis and prepuce.
  - o palpation and ultrasonography of the testes and epididymis to exclude these sources of inflammatory cells.
- transrectal examination and ultrasonography may demonstrate:
  - o enlarged, firm, and painful vesicular gland(s).
  - o significant variation exists between different stallions and within stallions following examinations at different times in:
    - ♦ size, degree of dilation, and ultrasonographic character of accessory glands.
    - ♦ findings of large, dilated glands on rectal examination and ultrasound do not confirm a diagnosis of vesiculitis.
- endoscopic examination of the urethra and colliculus seminalis is useful:
  - o gland opening may be inflamed, and purulent exudate seen draining from it in vesiculitis.
  - o confirm the diagnosis by passing flexible 5 French polyethylene catheter into the vesicular gland opening, via endoscopy, and aspirate fluid for cytological and bacteriological examination.

## Differential diagnosis

- bacterial culture from the semen alone does not confirm vesiculitis, as surface colonisation of the penis is the most common source of bacteria cultured in semen.
- other conditions causing PMN cells in the semen should be considered:
  - o orchitis/epididymitis, cystitis, or urethritis.

## Management

- vesiculitis is difficult to treat, and infections can become recurrent or chronic.
- appropriate systemic antibiotic therapy chosen based on sensitivity results.
- systemic anti-inflammatories to reduce inflammation and pain.
- combining systemic therapy with local therapy is more likely to be successful:
  - o vesicular gland is lavaged, via the endoscope and a flexible catheter.
  - o post lavage, appropriate non-irritating antibiotics are instilled into the gland.
- periodic semen evaluation and culture are used to monitor for recurrence.
- management of a stallion with chronic, low-grade infection can be achieved with the use of appropriate antibiotic-containing semen extenders:
  - o expose semen to the extender for at least 1 hour prior to insemination.

## Prognosis

- guarded for complete resolution.

## VENEREAL DISEASES

## Dourine (see also page 83)

## Definition/overview

- reportable to the World Organisation for Animal Health (OIE).
- North America, Australia, and New Zealand are dourine free.
- small number of cases are reported in western Asia, southeastern Europe, southern Africa, and Central America.
- true prevalence is unknown.

## Aetiology/pathophysiology

- venereal disease caused by the parasite *Trypanosoma equiperdum*.

## Clinical presentation and Diagnosis

- clinical presentation varies with the breed and the overall health status of the horse:
  - o locally adapted breeds and donkeys may be asymptomatically infected.
- clinical disease is characterised by:

- o mucopurulent urethral discharge, genital oedema, and fever.
- o followed by emaciation, hindlimb incoordination, and penile paralysis.
- o conjunctivitis and classical cutaneous plaques of 6–8 cm in diameter may occur.
- *T. equiperdum* can be identified in urethral exudates and buffy coat of blood samples:
  - o false negatives occur, particularly in advanced cases.
- complement fixation is the most reliable serological test:
  - o false positives occur due to cross-reaction with *T. brucei* and *T. evansi*.
- disease may be slowly progressive over weeks to years but is typically fatal.

## Differential diagnosis

- other causes of urethral discharge and penile paralysis.

## Management

- treatment with quinapyramin sulphate may eliminate clinical signs.
- in most countries, infected animals are euthanased as part of eradication programmes.

## Prognosis

- poor once clinical signs are evident.
- euthanasia is warranted to prevent transmission.

## Contagious equine metritis (CEM) (see also page 88)

## Definition/overview

- highly contagious venereal disease that occurs sporadically in Europe and many countries.
- thought to be eradicated from North America but sporadic outbreaks occur.
- control programmes are in place to prevent the introduction of CEM to disease-free countries and to eradicate the disease from Europe.

## Aetiology/pathophysiology

- caused by bacterium *Taylorella equigenitalis*:

- o fastidious microaerophilic gram-negative organism.
- transmission may occur through breeding, AI, or contact with contaminated equipment and fomites.

## Clinical presentation and Diagnosis

- stallions are asymptomatic carriers:
  - o positive cases may only be found during regulatory testing for export.
- mares bred to carrier stallions develop mucopurulent vaginal discharge caused by cervicitis and endometritis:
  - o begins 7–10 days after breeding.
  - o demonstrate a shortened inter-oestrus interval.
- follow government requirements and regulations for testing and reporting when CEM is suspected.
- diagnosis is made by isolating the bacterium from:
  - o urethra, urethral fossa, prepuce, or pre-ejaculatory fluid of stallions.
  - o isolation of the bacterium from the reproductive tract of mares after breeding to a suspect stallion.
- fastidious nature of the causative organism:
  - o swabs immediately placed in Ames' medium with charcoal and refrigerated during transport.
  - o if delayed, swabs may be frozen.
  - o culture requires a high $CO_2$ environment and selective media.
  - o usually performed at regulatory laboratories.
- serological testing is not useful in stallions because they do not mount an immune response.

## Differential diagnosis

- primary differential is acute bacterial endometritis of non-venereal causes:
  - o such as *E. Coli* or *Streptococcus* spp.
  - o venereal infections causing endometritis, including *Pseudomonas* and *Klebsiella*.

## Management

- follow government regulations regarding treatment and follow-up testing.
- CEM-positive cases are treated by:

- thoroughly washing the extended penis (teased to penile erection), prepuce, and urethral fossa in 2% chlorhexidine solution, taking care to remove all smegma.
- followed by packing with 0.2% nitrofurazone dressing on 5 consecutive days.
  - nitrofurazone ointment may not be available for veterinary use in some parts of the world.
- Silver sulphadiazine may be used as an alternative.
- USA and Canada require quarantine and testing, including the breeding and subsequent culture of 3 mares, for CEM for all stallions entering these countries.

## Prognosis

- infection in stallions is usually eliminated by treatment:
  - more than one cycle of treatment may be required.
  - some stallions may develop sensitivity to chlorhexidine, leading to soreness.
    - do not exceed the recommended concentration.

## Bacterial colonisation of the penis

### Definition/overview

- iatrogenic bacterial overgrowth can occur on the genital organs of:
  - intensively managed breeding stallions.
  - where there is excessive cleaning of the penis in colts and geldings (Fig. 2.42).

### Aetiology/pathophysiology

- external genitalia of the stallion normally harbour a mixed population of commensal bacteria.
- disruption of the normal bacterial population leads to:
  - heavy growth of potentially pathogenic species such as *Pseudomonas aeruginosa* and *Klebsiella pneumoniae*.
  - *Pseudomonas* is classified by serotype, phage type, and ability to cause haemolysis.

- haemolytic strains are considered infective.
- housing stallions on poorly maintained shavings encourages overgrowth of *Klebsiella*.
  - classified by capsular type; types 1, 2, 5, and possibly 7.
  - can be transmitted venereally and are associated with endometritis in the mare.
  - colonises itself on the external skin of the penis and prepuce.

## Clinical presentation and Diagnosis

- mares bred to the stallion have a lower conception rate than expected:
  - may return to heat sooner than 21 days due to endometritis.
  - uterine culture of multiple mares bred to the stallion results in growth of an identical microorganism.
- culture of the penis, prepuce, and urethral fossa before washing is required:
  - teasing the stallion to achieve an erection facilitates culture.
  - after washing and drying the penis, pre- and post-ejaculatory fluid is cultured.
  - semen is examined for WBCs to ensure that infection has not ascended to the accessory sex glands.

## Differential diagnosis

- other causes of subfertility.

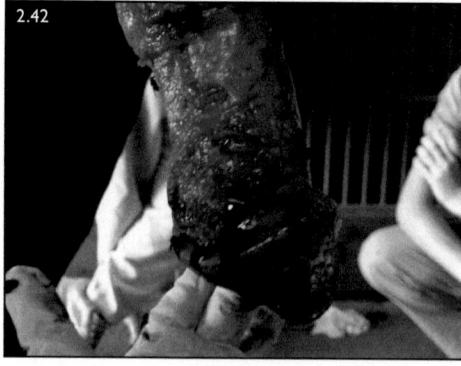

FIG. 2.42 Balanoposthitis in a 22-year-old gelding caused by frequent washing of the penis with harsh antibacterial soaps by the owners.

## Management

- prevent bacterial overgrowth by not using antiseptic cleansers when washing the penis:
  - clean, warm water and clean cotton for routine pre-breeding washing.
- *Pseudomonas* and *Klebsiella* infections can be resistant to treatment:
  - *Pseudomonas* infections can be treated by daily washing of the penis in a dilute solution of hydrochloric acid (2.5 ml 38% HCl/l water).
    - after washing and removal of smegma, some success has been reported with the use of a 1% silver nitrate solution spray daily for up to 30 days.
    - periodic treatment with sterile petroleum jelly helps prevent the penile skin from cracking during treatment.
  - *Klebsiella* infections can be treated by daily washing of the penis with a dilute hypochlorite solution (9 ml 5.25% Na hypochlorite/l).
    - Gentamicin solution (50 mg/ml) or ointment can be massaged into the penile and preputial skin, the urethral fossa, and the diverticulum following sodium hypochlorite washing (5.25%).
    - **Gentamicin ointment no longer available for veterinary use in Europe.**
- Enrofloxacin or other systemic antibiotics are used only if accessory sex gland infections are confirmed.
- progress should be monitored by follow-up swab samples collected every 2 days from:
  - urethra, urethral fossa and diverticulum, and preputial smegma 7 days after the end of treatment.
  - three sets of negative swabs can be assumed to indicate successful treatment.
- ongoing management includes:
  - thorough disinfection to prevent colonisation of equipment including buckets and AVs.
    - disposable plastic bag liner for wash buckets reduces cross-contamination.
  - infected stallions should have dedicated equipment to prevent pathogen transmission to other stallions.
- some cases can be managed by minimum-contamination breeding techniques and incubating semen in an extender containing the appropriate antibiotic for at least 1 hour prior to insemination.

## Prognosis

- good for treating colonisation of the external genitalia.

# Equine coital exanthema

## Definition/overview

- self-limiting venereal disease affecting mares (See page 81) and stallions.
- worldwide distribution.

## Aetiology/pathophysiology

- highly contagious disease caused by EHV-3 and spread through coitus or contaminated veterinary equipment.

## Clinical presentation and Diagnosis

- symptoms develop 4–8 days after infection:
  - 2 mm diameter red nodules appearing on the penis and preputial skin.
  - develop into vesicles and then pustules, which rupture, leaving shallow erosions (Fig. 2.43).
- fertility is not usually impaired, but stallions are reluctant to copulate due to pain.

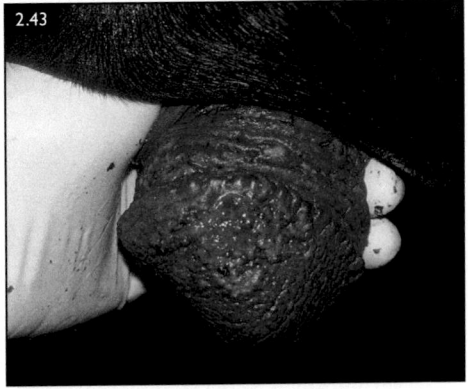

**FIG. 2.43** Acute coital exanthema ulcerative lesions on the glans penis of a stallion.

- course of uncomplicated infection is 3–4 weeks:
  - secondary bacterial infection of the lesions may occur.
  - unpigmented scars remain visible after recovery.
- usually diagnosed based on clinical presentation, including characteristic lesions.
- biopsy samples of lesions may be submitted for PCR testing or histopathology.

## Differential diagnosis

- trauma
- contact sensitivity
- bacterial infection
- SCC.

## Management

- strict sexual rest is required until the lesions are fully healed.
- **care must be taken to prevent iatrogenic transmission.**
- secondary bacterial infections of lesions may occur:
  - treated with appropriate topical medications.

## Prognosis

- excellent as the disease is self-limiting and resolves in 3–4 weeks.

# Equine viral arteritis (EVA)

## Definition/overview

- equine arteritis virus, causes EVA.
- stallions may become permanent carriers.
- worldwide distribution, but outbreaks are rare.
  - EVA is endemic in Standardbreds in many countries.
  - seroprevalence in Warmblood stallions is very high in many European countries.

## Aetiology/pathophysiology

- EVA is readily transmitted by respiratory and venereal routes:
  - venereal route is via acutely and chronically infected stallions shedding equine arteritis virus in the sperm-rich fraction of the ejaculate.

- mares acutely infected by either route usually seroconvert within 28 days and shed virus in body secretions.
- EVA also transmitted via embryo transfer (infected embryo donor mare to recipient).
- reason some stallions fail to clear the infection and become chronic carriers of the virus within the ampullary glands is not well understood but may be due genetic differences.

## Clinical presentation and Diagnosis

- asymptomatic infection is common in stallions, geldings, and non-pregnant mares.
- symptomatic cases, signs last 1–10 days and include:
  - fever, depression, anorexia, limb and scrotal oedema.
  - conjunctivitis and lacrimation, nasal and ocular discharge, and skin rash.
- CBC reveals a leucopaenia.
- scrotal oedema and fever may cause a decline in semen quality in stallions.
- abortion of a partially autolysed foetus occurs at any stage of gestation with little to no premonitory signs in pregnant mares.
  - mares infected in late gestation give birth to weak foals exhibiting a severe interstitial pneumonia.
- detection of the virus in farms experiencing clinical respiratory illness and abortion can be achieved using serology, virus isolation or PCR:
  - submission of aborted foetuses and membranes assists in confirming the diagnosis.
- laboratory confirmation of a diagnosis assists the clinician in providing advice as to appropriate vaccination and control measures.
- diagnosis of the persistent shedding state in the stallion is made via detection of equine arteritis virus in the semen by PCR testing or viral isolation.

## Differential diagnosis

- other respiratory infections, including influenza, EHV-1 or -4, equine adenovirus.

- other causes of abortion should be considered including *Leptospira*, and EHV-1 or 4.

## Management

- acutely infected animals should be treated with appropriate supportive care.
- no treatment is available to eliminate shedding in chronically infected stallions, other than castration.
- stallions should be vaccinated to prevent infection:
  - prior to vaccination, test for antibodies to equine arteritis virus.
  - serological testing cannot differentiate between natural infection and a vaccinate:
    - pre-vaccination titres documented before vaccination of colts to allow international transport.

- mares bred to infected stallions should be vaccinated 3 weeks prior to breeding:
  - must be isolated from pregnant mares for at least 21 days after vaccination.
  - modified live vaccine is available in North America.
  - only a killed vaccine is available in Europe.
  - vaccines should be used according to manufacturers' recommendation.
- EVA positive semen stallions can be used for breeding purposes provided:
  - strict isolation practices are upheld.
  - strict hygiene in the collection room is followed to prevent cross-contamination.
  - only bred to vaccinated mares.

## Prognosis

- poor in carrier stallions, as infection cannot be eliminated.

---

## SKIN DISEASE OF THE EXTERNAL GENITALIA

### Habronemiasis

#### Definition/overview

- granulomatous dermatitis caused by hypersensitivity to the larvae of *Habronema* spp.

#### Aetiology/pathophysiology

- adult *Habronema* spp. worms live in the equine stomach and eggs passed in the faeces are ingested by fly maggots in the manure.
- larvae are transferred to the skin by flies feeding around the mouth, eyes, male genitals, and lacerations.
- larvae deposited around the mouth are swallowed and develop into adults in the stomach.
- larvae deposited in other regions burrow into the skin, causing a granulomatous dermatitis.
- frequency of Habronemiasis has significantly reduced due to widespread use of ivermectin and moxidectin.

### Clinical presentation and Diagnosis

- occurs sporadically in a group of horses, but it can occur repeatedly in the same individual:
  - light-coloured horses are most commonly affected.
- occurs during warm weather, when fly populations are high.
- lesions are pruritic and appear as 1–25 cm reddish open sores, containing characteristic hard yellow granules.
- in stallions, blood noted in the genital region after mating may be the first detected sign:
  - proliferative lesions of the urethral process may interfere with urination and ejaculation.
  - inflammation and swelling of the genital region may disrupt spermatogenesis.
  - chronic, severe lesions of the prepuce may result in scarring and adhesion, interfering with normal penile erection.

- diagnosis based on history and clinical signs, particularly presence of yellow granules.
  - confirmed by biopsies.

## Differential diagnosis

- Sarcoid  • SCC  • granulation tissue.

## Management

- reduce incidence by maintaining a low parasite burden on breeding farms through regular deworming.

- limit exposure of susceptible stallions to flies via appropriate housing, manure management, and use of insecticides.
- systemic ivermectin is used to kill larvae in clinical cases:
  - lesions are surgically debulked.
  - treated with application of steroid cream to reduce inflammation.

## Prognosis

- good with appropriate management.

## TUMOURS OF THE MALE REPRODUCTIVE TRACT

# Tumours of the Penis and prepuce

## Squamous cell carcinoma (SCC)

### Definition/overview

- most common neoplasm affecting the skin of the penis and prepuce.

### Aetiology/pathophysiology

- most common in animals with non-pigmented skin on the penis and prepuce, such as Appaloosas and Pintos.
- geldings are more frequently affected than stallions:
  - possibly more smegma accumulation and penis is less frequently washed.

  - carcinogenic action of smegma accumulation on the penis has been suggested.
- association between equine papilloma virus EcPV2 and SCC has been found.

### Clinical presentation and Diagnosis

- non-healing erosions/small growths may be noted on the skin during washing for breeding.
- in non-breeding stallions or geldings early on.
  - swelling of the sheath, reluctance to exteriorise the penis, and foul odour.
- most affected areas:
  - internal prepuce (Fig. 2.44), glans penis (Fig. 2.45), and urethral opening.
- two presentations are common:

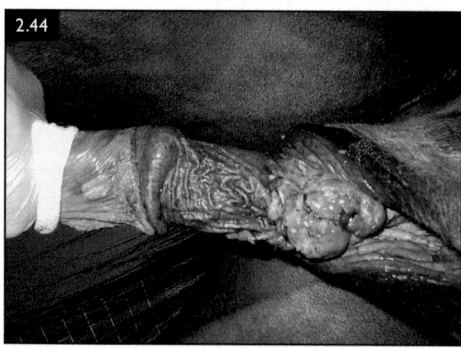

FIG. 2.44 SCC lesion at the base of the penis of a stallion.

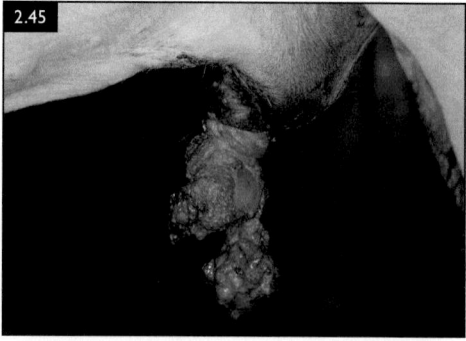

FIG. 2.45 A severe erosive and granulating SCC of the glans and body of the penis of an aged gelding.

- proliferative form (cauliflower-like growths) that may ulcerate and bleed.
- invasive form that destroys penile architecture and distorts the penis and prepuce.
- precancerous skin changes can include:
  - discrete areas of depigmentation and thickening of the penile skin (See Fig. 2.10).
- penis must be fully exteriorised to permit examination of its entirety, including examination of the fossa glandis, and palpation of the penile base.
- rectal examination of superficial inguinal and pelvic lymph nodes may indicate metastasis.
- diagnosis is made by histopathology of a biopsy.

## Differential diagnosis

- habronemiasis.
- other skin tumours such as sarcoids, melanoma, and viral papilloma.

## Management

- small SCC lesions can be treated by cryosurgery:
  - liquid nitrogen or carbon dioxide delivered through a cryoprobe or liquid nitrogen spray.
  - lesions are frozen to a depth of 3 mm, with three freeze–thaw cycles.
- small lesions can be treated by repeated application of 5% 5-flourouracil every 2 weeks.
- intralesional injection of bleomycin or mitomycin has been reported.
- possible negative effects on spermatogenesis of cytotoxic chemotherapeutic agents should be considered before commencing therapy in a breeding stallion.
- more extensive lesions may be treated by circumferential reefing surgery.
- advanced cases:
  - castration followed 3 weeks later by penile amputation (Figs. 2.46, 2.47).

## Prognosis

- can be refractory and difficult to treat, although slow to metastasise.

# Other tumours of the penis and prepuce

## Definition/overview

- variety of cutaneous neoplasms may affect the skin of the penis and prepuce.

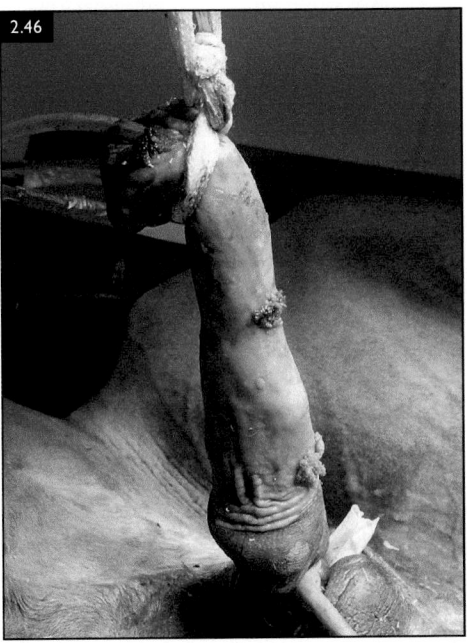

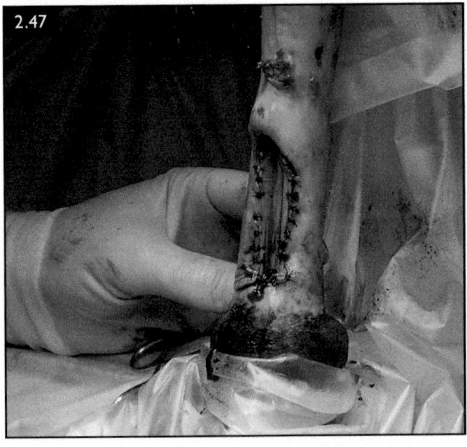

**FIGS. 2.46, 2.47** Immediately preoperative view (2.46) of the penis of a gelding showing SCC lesions at two sites on the body and a larger lesion on the glans. The gelding is undergoing a penile amputation (2.47) and at this stage the urethrotomy incision has just been completed.

## Sarcoids

- fibroblastic tumours of the skin.
- most common in younger horses.
- fibroblastic nodular (most common), verrucous (warty), mixed, or occult (rare) type (Fig. 2.48).
- preputial and penile sarcoids are managed similarly to sarcoids in other locations:
  - surgical excision is likely to be the best option, but recurrences are common.
  - cryotherapy reduces recurrence:
    - used with caution on penile, preputial, and scrotal skin due to the potential for scarring and fibrosis.
    - may alter function (penis and underlying urethra) and fertility (scrotum).

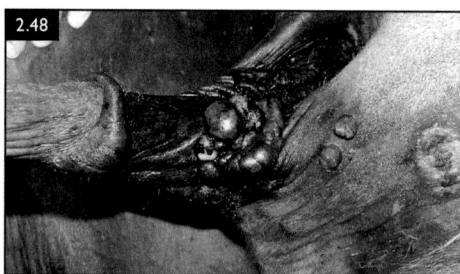

**FIG. 2.48** View of a horse with sarcoid lesions at the base of the prepuce (nodular fibroblastic) and also on the adjacent skin.

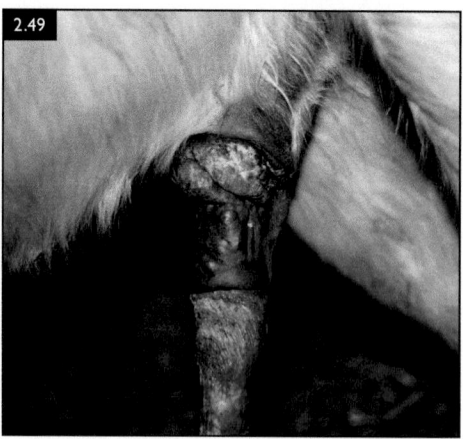

**FIG. 2.49** Small melanomas of the base of the penis of an aged grey gelding pony. They were not associated at this stage with any clinical signs.

## Melanomas

- involving the prepuce or penis of older geldings and stallions (Fig. 2.49).
- in other locations, such as the perianal or facial regions.
- usually <3 cm, multiple, smooth, round, and hairless.
- slow to metastasise.
- oral cimetidine (2.5 mg/kg q24h) may be helpful in slowing progression.

## Viral papillomas

- tend to affect young colts or geldings.
- resolve spontaneously over 4–12 weeks without treatment.
- other lesions are usually found on the muzzle of affected animals.

# Testicular tumours

## Definition/overview

- include seminomas, Leydig cell tumours, Sertoli cell tumours, and rarely others such as teratomas, lipomas and fibromas.

## Aetiology/pathophysiology

- Seminomas are the most common testicular tumour:
  - arise from the germinal epithelium of the seminiferous tubule:
  - most common in older stallions and are grey and lobulated (Fig. 2.50).
- Interstitial tumours are rare and arise from the testosterone-producing Leydig cells:
  - tan, firm, and nodular.
- Sertoli cell tumours are very rare, occurring primarily in retained testes:
  - firm and grey-white.
- Teratomas are rare and occur primarily in retained cryptorchid testes:
  - may be large and contain various tissues, including bone and hair.

## Clinical presentation

- Seminomas cause testicular enlargement over time in older stallions:
  - affected testis will feel enlarged and lobular/lumpy on palpation.
  - normal testicular tissue is replaced with locally invasive seminoma.
  - metastatic disease is possible.

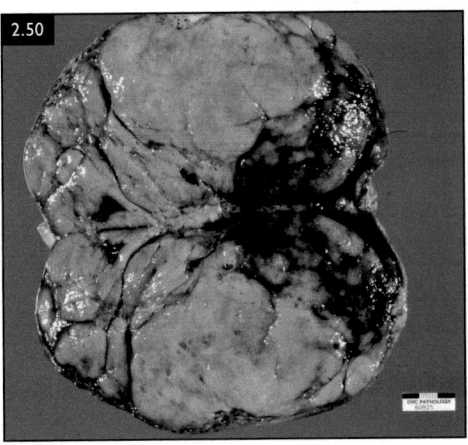

**FIG. 2.50** Gross cut section of the testis of stallion with a seminoma which has invaded most of the testis, replacing normal tissue with tumour.

- Interstitial cell or Leydig cell tumours, are rare by comparison.
  - more commonly found in a retained testicle and are usually benign.
- Leydig cell tumours are firmer and more nodular on palpation than seminomas.
- Sertoli cell tumours are quite rare in stallions.

## Differential diagnosis

- testicular torsion
- abscess
- granuloma
- trauma.

## Diagnosis

- palpation and ultrasound (7.5 MHz) comparison with the normal testicle (Fig. 2.51).

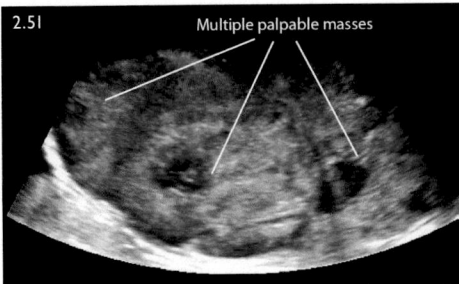

Multiple palpable masses

**FIG. 2.51** Ultrasound appearance of testicle with a seminoma in 24-year-old stallion with onset of infertility and poor semen quality over the previous 6 months.

- cryptorchid testes can be examined by rectal palpation and ultrasonography prior to surgery.
- tumour type is confirmed by testicular biopsy or, more commonly, by histopathology after castration.
- thickening of the spermatic cord, enlargement of the pelvic and abdominal lymph nodes, or histopathological detection of tumour cells in the cord indicates metastasis.

## Management

- unilateral castration of the affected testis:
  - spermatic cord should be transected as far proximal as possible.
  - ligation may be necessary to control haemorrhage.
- cryptorchid stallions, bilateral castration should be performed.

## Prognosis

- usually are benign.
- return to fertility may be compromised in stallions with long-standing tumours due to:
  - old age.
  - hormonal down-regulation.
  - inflammation of the normal testicle secondary to scrotal oedema or unilateral castration.

# Tumours of the spermatic cord

## Definition/overview

- spermatic cord tumours are extremely rare in stallions.
- fibroma, leiomyoma arising from the wall of cord vessels, and mesothelioma of the tunica albuginea have been reported.

## Clinical presentation and Diagnosis

- may present with scrotal pain, reluctance to breed, generalised depression, and ejaculatory pain.
- some cases may have no outward signs.
- palpation of the spermatic cord reveals a mass.

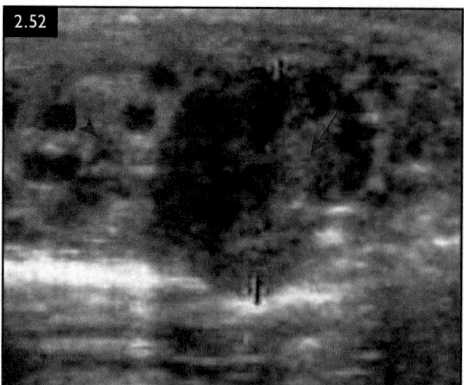

FIG. 2.52 Ultrasound image of a hypoechoic, firm mass (red arrow) found in the left spermatic cord (yellow arrow) of a 20-year-old Hanoverian stallion that presented for painful ejaculation.

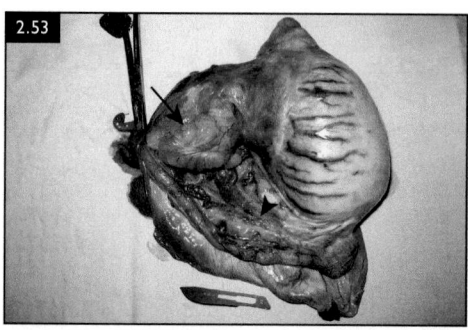

FIG. 2.53 Gross appearance of the testis from the stallion in 2.52 following unilateral castration. The mass (red arrow) was within the spermatic cord (yellow arrow), with adhesions to the tunic. The histological diagnosis was a leiomyoma.

- ultrasonographic evaluation may be suggestive of neoplasia (Fig. 2.52).
- diagnosis can be confirmed histologically by biopsy of the mass.

## Differential diagnosis

- thrombosis of spermatic cord vessels
- varicocoele.

## Management

- unilateral castration with removal of as much of the spermatic cord as possible on the affected side (Fig. 2.53).

## Prognosis

- for life, recovery, and return to fertility is dependent on the type of tumour and whether metastasis has occurred.

# EQUINE CASTRATION

- most common routine surgical procedure performed in the horse.
- consists of removal of one or, more commonly, both testicles.
- routine procedure and normally performed as an elective surgery:
  - complications are relatively common.
  - knowledge of urogenital anatomy and preparation for the procedure greatly reduces the complication rate.

## Indications for castration

- most castrations are intended to suppress the development of stallion-like characteristics.
- age at which the surgery is performed:
  - first 2 months of life is not advisable, as young foals are more vulnerable to stress and infections.

  - colts over 3 months old can be castrated very successfully, particularly when still suckling.
  - more commonly, colts are castrated several months after weaning or during the first year of life.
  - competition horses may be kept entire until a later stage to assess their performance and breeding potential.
  - more a horse is allowed to develop into sexual maturity, the less likely it is that male characteristics will be suppressed.
- other indications for castration include:
  - sterilisation of a male no longer intended for breeding purposes.
  - salvage procedure in cases of:
    - severe trauma to the testicles or scrotum.

♦ testicular neoplasia, intractable infection or a chronically inflamed testis.
  – unilateral castration if possible to maintain breeding soundness.
- generally accepted that cryptorchids should be castrated:
  ○ evidence of hereditary predisposition.
  ○ breed societies will not allow registration of cryptorchid stallions.

## Postoperative complications and their management

### Haemorrhage (Fig. 2.54)

- after surgery, a small amount of bleeding is expected:
  ○ usually originates from subcutaneous vessels and stops within a few hours.
  ○ stream of blood persisting for more than 15 minutes considered excessive.
- decision to intervene made as early as possible to minimise blood loss.
- horse should be sedated, and an attempt made to find the origin of the haemorrhage.
- surgical exploration under general anaesthesia may be required.
- less severe cases:
  ○ packing of the scrotum with sterile swabs may stop smaller haemorrhages.

○ testicular stump is bleeding:
  ♦ re-emasculate and ligature.
- patient should be given systemic antibiotics for at least 5 days.

## Swelling

- minimal postoperative swelling is expected in the first 24–48 hours after surgery:
  ○ usually peaks at 4–5 days, particularly in closed castrations.
- excessive swelling should be investigated.
- open castrations (Fig. 2.55) most common cause of swelling is:
  ○ premature wound closure.
  ○ incorrect positioning of the incision causing drainage impairment.
- castration wound should be examined, opened and palpated by sterile digital exploration.
- increasing exercise may reopen the wounds, thus avoiding further manipulation of the surgical sites.
- closed castrations with swellings may be harder to investigate:
  ○ ultrasonographic imaging may help identify the cause of the problem.
  ○ bleeding, eventration and infection can all cause swelling and must be differentiated.
  ○ following closed inguinal castration, air can get trapped in the subcutaneous fascial planes during surgery and accumulates in the scrotum.
    ♦ no clinical concern although possibly alarming to the owner.

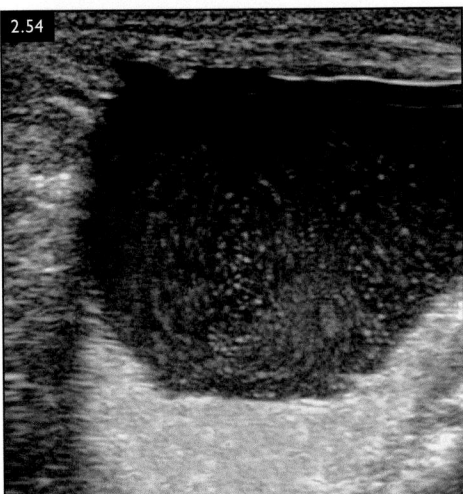

**FIG. 2.54** Ultrasonographic image of a recently castrated horse with excessive postoperative bleeding, revealing a large collection of blood trapped within the scrotum.

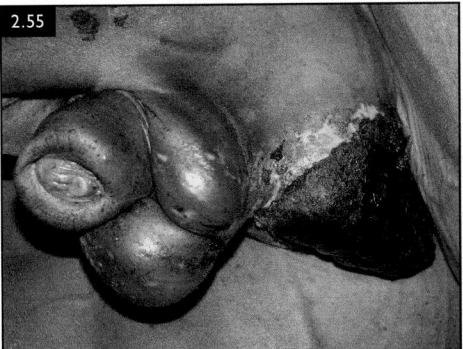

**FIG. 2.55** This horse had excessive swelling post castration, which has led to ischaemic necrosis of the scrotum. Note the residual swelling of the end of the prepuce.

## Infection

- reported incidence of postoperative infection after castration varies (5% to 20%).
- strict aseptic surgery is difficult to achieve in field situations.
- ascending infection can also occur when the wounds are left open.
- most infections respond well to treatment but can become serious if left untreated.

### Localised infection

- majority of infections are localised to the scrotal and/or inguinal tissues.
- mainly caused by impaired drainage allowing bacteria to multiply within the dead space.
- most cases presented with:
  - obvious swelling.
  - and/or change in discharge appearance and smell (e.g. purulent).
- systemic signs may be subtle in the early stages:
  - depression accompanied by a reduced appetite and pyrexia.
  - haematological parameters vary depending on the severity and time elapsed since the beginning of the infection process.
- reopening of surgical wounds is indicated when premature closure has occurred.
- antibiotics may be indicated in more severe cases.
- where primary closure was performed:
  - swelling can be more prominent and systemic effects more severe:
    - drainage is not possible.
    - establishing drainage by opening the scrotum is essential.

### Peritonitis

- cellular inflammatory response in the peritoneal cavity occurs after routine castration.
- more severe response occurs if infection tracks up the inguinal canal and causes a septic peritonitis.
- associated endotoxaemia will cause related clinical signs.
- colic and inappetence may be present.
- analysis of peritoneal fluid through abdominocentesis will confirm the diagnosis.

- treatment should include establishment of drainage, aggressive broad-spectrum antibiotic therapy and anti-endotoxaemia therapy.

## Eventration

- eventration means that abdominal viscera have prolapsed through the inguinal ring.
- small intestine is more common than omentum.
- prompt management of the situation should give a moderate to good prognosis.
- usually occurs in the first few hours after surgery (Fig. 2.56).
- patient should be sedated, and the prolapsed tissue palpated and anatomically identified.
- small intestine should be lavaged and kept moist with a wet cloth until, under GA, the portion of gut is placed into the scrotum and the scrotal wounds sutured closed.
- horse should be administered broad-spectrum antibiotics and referred to a surgical facility:
  - replacement of the small intestine usually requires a coeliotomy (Fig. 2.57).

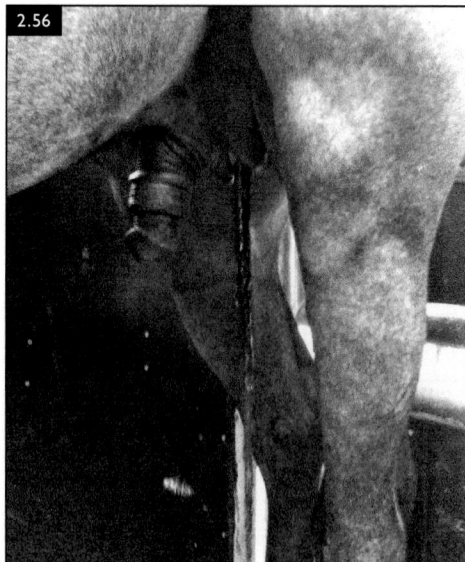

**FIG. 2.56** Eventration of omentum in a horse castrated by a standing open procedure earlier in the day.

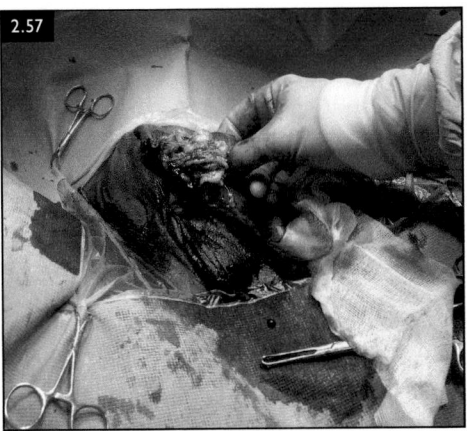

FIG. 2.57 This 6-month-old Warmblood was castrated by the open method 24 hours prior to this picture. Investigation under general anaesthesia in dorsal recumbency has confirmed eventrated omentum from the left vaginal tunic. This was resected and replaced into the abdomen. The superficial inguinal ring and vaginal tunic were then closed.

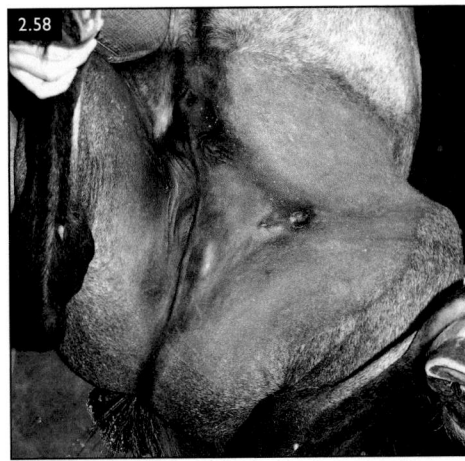

FIG. 2.58 A horse under GA and in dorsal recumbency prior to preparation for surgery to remove a scirrhous cord. Note the chronically draining tract hole in the cranial left side of the inguinal region.

- if omentum is involved, the same procedure can be performed:
  - where financial restraints are involved, some patients recover after omentectomy performed as proximal as possible using emasculators.

## Funiculitis

- infection of the emasculated stump:
  - if chronic and suppurative, the term 'scirrhous cord' is often used.
  - 'Champignon' and 'botryomycosis' also terms used to describe this condition.
- most common bacterial isolates are gram-positive bacteria:
  - *Staphylococcus* spp. and *Streptococcus* spp.
- risk of funiculitis may be increased if ligatures are placed during the castration:
- break in sterility during surgery is probably more likely for infection to occur.
- contamination of the emasculators is a common risk factor.
- failing to resect the vaginal tunic and external cremaster muscle may also increase the risk.
- rarely, the infected tissue may extend proximally into the peritoneum.

- history may reveal the surgical site healed without complication:
  - subsequently, a draining tract occurred.
  - alternatively, a small draining tract may have remained after surgery (Fig. 2.58).
  - initial clinical signs are swelling in the area and drainage near the surgical scar.
  - discomfort in the inguinal area or hindlimb lameness may be present.
- condition can remain dormant and clinically silent for several months or even years.
- clinical examination reveals a thickened cord palpable within the inguinal region (Fig. 2.59).
- ultrasonography will help determine the extent of the infected tissue.
- antibiotic therapy provides temporary improvement, but once discontinued, clinical signs usually return.
- treatment of choice for most cases is surgical and involves:
  - reopening of the skin around the draining point and careful soft-tissue dissection around the draining tract (See Fig. 2.33).
  - all infected spermatic cord must be removed by emasculation of the cord proximal to any thickened tissue.

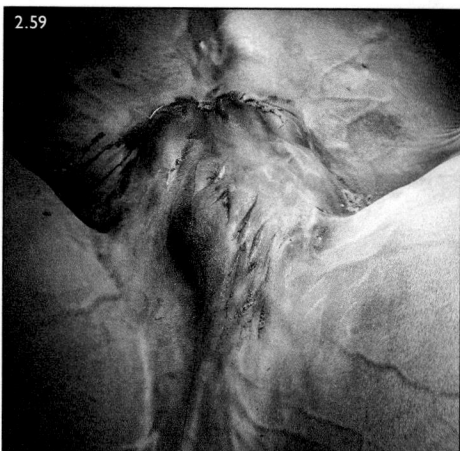

FIG 2.59 Note the swelling on the left side of the prepuce and inguinal region due to a scirrhous cord.

- ○ ligatures are not placed, and the surgical site is left open to heal by secondary intention.
- ○ systemic antibiotics and NSAIDs are usually required for 5–10 days.
- ○ drainage is encouraged by regular in-hand walking.

## Hydrocoele

- collection of peritoneal fluid in the scrotal region following castration due to persistent communication at the end of the emasculated vaginal tunic.
- can occur weeks to months following the surgery.
- usually of no clinical consequence.
- diagnosis is made by palpation of the swelling:
  - ○ reveals fluid that can be pushed out of the scrotum and into the abdominal cavity.
- ultrasonography may help to rule out the presence of any other soft tissue problems.

- treatment is not normally required except for cosmetic reasons.
- surgical treatment requires reopening of the castration site and re-emasculation of the stump.

## Penile damage

- occurs if the shaft of the penis is confused with one of the testes during surgery.
- iatrogenic damage occurs.
- careful pre- and intraoperative palpation should be performed, to help avoid this.
- treatment depends on the structures damaged:
  - ○ haemostasis is essential.
  - ○ urethra not been damaged, the wound may be sutured.
  - ○ extensive damage caused, penile amputation or retroversion may need to be performed as a salvage procedure.

## Continued masculine behaviour

- persistent masculine behaviour may be seen in up to 30% of cases following castration.
- reasons are not fully understood:
  - ○ likely to involve innate group behaviour between horses.
  - ○ hormonal influence may be involved:
    - ♦ androgens are not produced solely by testicular tissue.
- cryptorchidism and incomplete castration should be ruled out by hormonal blood tests.
- possibility of continued masculine behaviour should be discussed with the owners before surgery:
  - ○ especially in older males and those used previously as stallions.
  - ○ risk of maintaining masculine behaviour is increased.

# The Foal

## Introduction

- early signs of disease can be non-specific.
- knowledge of normal behaviour patterns and physiological parameters is essential (Table 3.1).
- systematic clinical approach ensures that subtle signs and potential complications are not overlooked.

## Examination of the Normal Neonatal Foal

### History

- detailed history is essential and should include:
  - mare's health and past breeding record.
  - current pregnancy, parturition and length of gestation.
  - history of vulval discharge and premature lactation.
  - placental examination findings.

### Behaviour

- should be well bonded to the mare and show inquisitive/active behaviour.
  - righting and suck reflexes within the first few minutes after birth.
  - stand within 1–2 hours and suckle from the mare within 2–3 hours.
- healthy foals typically feed 5–7 times an hour and pass large volumes of dilute urine.

## Physical examination

- thorough physical examination should be carried out.
- physical appearance:
  - look for signs of immaturity and intrauterine growth retardation:
    - ◆ low body weight, poor body condition and generalised weakness.
    - ◆ floppy ears, mole-like coat and domed forehead.
  - assess for congenital deformities.
  - head examined for wry nose, parrot mouth, facial swellings or milk at nostrils.
  - mucous membranes examined for:
    - ◆ scleral haemorrhage or congestion may indicate a traumatic birth.
    - ◆ jaundice – neonatal isoerythrolysis or Equine Herpesvirus infection.
    - ◆ petechiation or congestion (+/- on the pinnae) can occur with sepsis.
    - ◆ dry membranes – inadequate feeding or dehydration.
  - examine the eyes:
    - ◆ entropion due to dehydration or idiopathic:
      - – can lead to corneal ulceration.
    - ◆ uveitis with bacteraemia/sepsis.
    - ◆ fundic haemorrhages:
      - – common in Thoroughbred foals and of no clinical significance.
- examine the respiratory tract:

**TABLE 3.1** Normal heart and respiratory rates and rectal temperature of the neonatal foal

| AGE | HEART RATE (BEATS/ MINUTE) | RESPIRATORY RATE (BEATS/MINUTE) | TEMPERATURE °C (°F) |
|---|---|---|---|
| 1 minute | 60–80 | Gasping | 37–39 (99–102) |
| 15 minutes | 120–160 | 40–60 | 37–39 (99–102) |
| 12 hours | 80–120 | 30–40 | 37–39 (99–102) |
| 24 hours | 80–100 | 30 | 37–39 (99–102) |

DOI: 10.1201/9781003386834-3

- ○ increased respiratory rate and effort may indicate pulmonary disease.
- ○ lung auscultation should reveal:
  - ◆ relatively loud bronchovesicular sounds in the normal foal.
- ○ abnormal chest contour or pain/crepitus on rib palpation indicates:
  - ◆ rib fractures.
    - – most common location ventral chest, just behind the elbows.
- • auscultate the heart:
  - ○ heart rate in the young foal is very labile (assess when relaxed).
  - ○ transient arrhythmias are common in newborn foals.
  - ○ grade I–IV left-sided holosystolic murmur often audible for the first 2–3 days:
    - ◆ associated with patent ductus arteriosus.
  - ○ flow murmur may be audible at the left heart base:
    - ◆ may persist for many days.
  - ○ signs of severe congenital cardiac disease include:
    - ◆ exercise intolerance, cyanosis and abnormalities in peripheral pulses.
    - ◆ cardiovascular compromise merits further investigation:
      - – including radiography and echocardiography.
- • abdomen evaluated for distension or a 'tucked up' appearance.
- • umbilical region and remnants palpated for thickening, dampness or herniation.
- • normal foals initially pass large quantities of dark brown meconium:
  - ○ followed by paler milk faeces.
- • limbs evaluated for joint effusion, and flexural and angular limb deformities.

## Physiological differences between the neonatal foal and the mature horse

- • newborn foals have a larger total body water content.
- • larger extracellular fluid and plasma volume, which reduces over the first 4 weeks.
- • particularly susceptible to water loss (high surface area to volume ratio).
- • equine neonatal kidney reduced capacity for sodium excretion:

- ○ lower tolerance of high sodium load.
- • hepatic function is immature for first 7–14 days of life:
  - ○ may slow drug metabolism.
- • differences in drug pharmacokinetics means:
  - ○ dose rates of some antibiotics are higher, others have shorter dosing interval.
  - ○ use foal-specific doses for drugs, especially antibiotics (Table 3.2).
  - ○ approximately 1 month of age drug metabolism becomes similar to adult.
  - ○ foals up to 5 months of age are more tolerant of oral antimicrobial administration due to their immature colon function.
- • neonatal foals are susceptible to disturbances in blood glucose and serum electrolyte levels:
  - ○ can occur with disrupted milk ingestion.
- • thermoregulatory function is less able to compensate for environmental temperature changes.
- • agammaglobulinaemic at birth and immunologically naïve:
  - ○ reduced levels of many of the components of the non-specific immune system.
  - ○ antibody levels and effective neutrophil function are dependent on adequate transfer of colostral immunity.

## Routine care of the neonatal foal

- • preventive medicine can reduce the incidence of neonatal disease and should include:
  - ○ good nutrition, worming, and careful monitoring of the mare in late pregnancy.
  - ○ good hygiene at foaling.
    - ◆ foaling box should be clean, disinfected between foalings, and well-bedded.
    - ◆ mare's udder and perineum cleaned with warm soapy water.
    - ◆ tail bandaged during foaling.
- • **ingestion of adequate quantities of good quality colostrum is essential:**
  - ○ routine checking of colostral quality is very helpful:

**TABLE 3.2** Dosages of some drugs used in the foal

| DRUG | ROUTE | FREQUENCY (HOURLY INTERVAL) | DOSE (MG/KG UNLESS OTHERWISE SPECIFIED) |
|---|---|---|---|
| **ANTIMICROBIALS** | | | |
| Amikacin | i/v | 24 | 20–25 if 2–4 weeks old<br>25–30 if <7 days old |
| Ampicillin | i/v | 6 | 20 |
| Cefotaxime | i/v | 6 | 50–100 |
| Ceftiofur | i/m | 12 | 5–10 |
| Clarithromycin | p/o | 12 | 7.5 |
| Doxycycline | p/o | 12 | 10 |
| Gentamicin sulphate | i/v | 24 | 6.6 if >7 days old<br>10 if <7 days old |
| Metronidazole | p/o | 12<br>12<br>8 | 10 if <7 days old<br>15 if 10–12 days old<br>15 if >2 weeks old |
| Oxytetracycline | i/v | 12 | 5–10 |
| Penicillin (Na or K) | i/v | 6 | 22,000 IU/kg |
| Rifampin | p/o | 12 | 5–10 |
| Trimethoprim sulphate | p/o | 12 | 30 |
| **ANALGESICS, SEDATIVES AND ANTI-INFLAMMATORIES** | | | |
| Butorphanol | i/v | As required | 0.01–0.1 |
| Diazepam | slow i/v | Single bolus | 0.1–0.2 |
| Meloxicam | i/v | 12 (<6 weeks)<br>24 (>6weeks) | 0.6 |
| **RESPIRATORY STIMULANTS** | | | |
| Caffeine | p/o | 24 | 10 loading, then 2.5–3 |
| **SEIZURE CONTROL** | | | |
| Midazolam | i/v<br><br>i/v | Constant rate infusion<br><br>Single dose | 0.02–0.06 mg/kg/h (persistent seizures)<br>0.04–0.1 (short term seizure control) |
| Phenobarbital | i/v<br>p/o | 8–24<br>12 | 2–10<br>3–10 (monitor serum concentrations) |
| Phenytoin | i/v | 6 | 5–10 initially,<br>1–5 maintenance |

♦ identifies poor quality in some mares and loss from premature lactation.
  – sugar refractometer reading of >25% indicates excellent quality.
  – <15% indicates very poor quality.

♦ foals that fail to nurse within 3 hours or those high risk for disease:
  – given good quality colostrum (minimum of 500 ml for a 50 kg foal) by nasogastric tube.

- routine blood samples taken at 12–36 hours can identify early signs of disease and efficacy of passive immunity transfer:
  ○ haematology, total proteins, immunoglobulin, and inflammatory proteins.

- umbilicus should be kept clean and dry:
  ○ 0.5% chlorhexidine solution for topical application:
  ○ directly after birth and then 2–3 times over the first 24 hours.

## IMMUNODEFICIENCIES

# Failure of transfer of passive maternally derived immunity (FTPI)

## Definition/overview

- most common immunodeficiency in the newborn foal.
- foals are agammaglobulinaemic at birth:
  ○ dependent on ingestion and absorption of adequate quantities of good quality colostrum soon after birth for the transfer of antibody.
- FTPI increases susceptibility to infection in the neonatal and paediatric periods.

## Aetiology/pathophysiology

- epitheliochorial equine placenta prevents transfer of maternally derived antibodies to the foal *in utero*.
- transfer of colostral immunoglobulins and other important immune factors (including complement, lysozyme, lactoferrin, and B lymphocytes) is essential.
- short-lived specialised cells in the foal's small intestine pinocytose large molecules, such as immunoglobulins, which then pass into the circulation.
- failure of this process results in low levels of circulating antibodies and compromise of the non-specific immune system, rendering the foal susceptible to infection.
- mare factors causing FTPI:
  ○ premature lactation and loss of colostrum is the most common cause.
  ○ poor quality colostrum is seen repeatably in some mares.
- foal factors causing FTPI:
  ○ failure to absorb IgG from ingested colostrum is uncommon, but possible:
    ♦ most commonly after 12–24 hours of age when the absorption process has closed.

  ○ failure to ingest adequate quantities of colostrum due to problems with:
    ♦ foal standing, and/or successfully nursing within the critical time frame.

## Clinical presentation

- detected on routine screening of blood samples at 12–36 hours of age.
- when investigating suspected infectious disease in newborn foals.

## Differential diagnoses

- Congenital immunodeficiencies.

## Diagnosis

- blood samples taken at 12–36 hours to measure serum IgG detect FTPI.
- serum IgG levels <4 g/l (0.4 g/dl) indicates FTPI:
  ○ between 4–8 g/l (0.4–0.8 g/dl) increase susceptibility to infection:
    ♦ particularly when the level of infectious challenge is high.
  ○ >8 g/l (>0.8 g/dl) is considered adequate.
- serum globulin levels >12 g/l (>1.2 g/dl) are likely to correlate with IgG levels >4 g/l (0.4 g/dl) in healthy foals (not reliable in high-risk foals).
- several different tests are available to measure IgG concentration:
  ○ large commercial laboratories commonly use immunoturbidometric assays.
  ○ point-of-care semi-quantitative tests for IgG (less accurate) include:
    ♦ ELISA SNAP test
    ♦ zinc sulphate turbidity test.
    ♦ glutaraldehyde coagulation test
    ♦ latex agglutination test.
    ♦ a commercial portable immunoturbidometric assay is now available.

## Management

- prevented during the first 4–6 hours after birth by administration of high-quality colostrum via bottle or stomach tube:
  - 1–2 litres are optimal but depends on colostrum availability and quality.
  - 300–500 ml can be fed every 1–2 hours.
  - oral plasma is an option but is expensive and less concentrated.
  - colostrum substitutes are available:
    - none contain other immune factors and not all contain equine IgG.
    - severe anaphylactic reactions can occur in foals that subsequently receive a plasma transfusion.
- plasma transfusion is the only treatment once there is closure of the specialised small intestinal transfer mechanism by 24 hours old:
  - commercially produced frozen hyperimmune plasma is available:
    - transfusion of 1 litre of commercial plasma to a 50 kg foal typically raises IgG concentration by 2g/l.
  - alternatively, fresh plasma may be harvested aseptically from a donor:
    - disease-screened, cross-matched, with high IgG concentration (>15 g/l).
  - administered via an aseptically placed 16-gauge jugular catheter and blood-giving set with an inline filter:
    - start very slowly, and monitor foal for signs of:
      - adverse reaction, cardiovascular overload, or anaphylaxis.
      - if tolerated, increase the rate gradually.
      - 1 litre administered over 20–30 minutes in a clinically healthy 50 kg foal to minimise stress.
      - if two litres are required, administer the second on a separate occasion or at a slower rate to prevent circulatory overload.
      - if adverse reaction signs are mild, it may be possible to continue the transfusion at a slower rate.
    - gentle manual restraint or the 'foal squeeze' should be used during plasma transfusion if possible.
      - chemical restraint may mask subtle transfusion reactions.
  - check IgG levels the following day to ensure satisfactory levels are reached.
  - prophylactic broad-spectrum antibiotics may be indicated depending on the specific situation of the foal, but their use is controversial.

## Severe combined immunodeficiency (SCID)

### Definition/overview

- most frequent in Arabians and occasionally Appaloosas.
- inherited autosomal recessive trait.

### Aetiology/pathophysiology

- homozygous foal cannot produce mature B and T lymphocytes:
  - heterozygous carriers of the gene are clinically normal.
- affected foals with effective transfer of colostral immunity usually healthy for the first 1–3 months of life.
- as maternally derived antibody levels decrease:
  - foal unable to produce endogenous antibody and cell-mediated immune responses, succumbing to recurrent infections.

### Clinical presentation

- foals are normal at birth and then develop recurrent severe infections:
  - often involve the respiratory system, within the first 2–3 months of life.
  - atypical organisms (adenovirus, *Pneumocystis jiroveci, E. coli*).
  - usually die by 6–9 months of age of bacterial pneumonia and/or enteritis.

### Differential diagnosis

- other immunodeficiencies such as selective IgM deficiency
- bacterial pneumonia.

### Diagnosis

- based on clinical signs, breed, and age of the foal.
- once clinical signs become apparent:
  - persistent lymphopaenia.

- absence of circulating IgM (after maternally derived levels have waned).
- lymphocytic hypoplasia of lymph nodes, spleen, and thymus:
  - seen at biopsy or post-mortem examination.
- diagnosis confirmed by genetic testing at specialist laboratories.

## Management

- no specific therapy.
- supportive therapy may extend life for a short period by treating opportunistic infections, but fatal infections are inevitable.

## Prognosis

- hopeless, and humane destruction is indicated once diagnosis confirmed.
- genetic testing can detect carrier animals, which should be removed from the breeding population.

# Foal immunodeficiency syndrome (Previously Fell Pony Syndrome)

## Definition/overview

- fatal immunodeficiency syndrome affecting Fell and Dales pony foals.
- characterised by severe anaemia and lymphopaenia.
- caused by a mutation on chromosome ECA26 with autosomal recessive inheritance.
- foals are healthy at birth, but usually present at 1–16 weeks of age with:
  - chronic, multiple, and severe viral/bacterial/fungal infections:
    - often respiratory or gastrointestinal.
  - foals usually die by 2–6 months of age.
  - oral candidiasis is common and severe skin infections also occur.
- severe non-regenerative anaemia, lymphopaenia, low IgM levels and peripheral ganglionopathy.
- diagnosis made via genetic testing.
- prognosis is hopeless and humane destruction is indicated.
- future breeding implications for the mare and stallion, should be considered.

# Immune-mediated thrombocytopaenia

See Book 5.

# Neonatal isoerythrolysis (NI)

## Definition/overview

- characterised by the destruction of the foal's RBCs due to the presence of alloantibodies of maternal origin that were absorbed from colostrum.
- occurs in newborn foals within the first week of life.

## Aetiology/pathophysiology

- 32 blood-group antigens in the horse:
  - 90% of cases associated with antigens Aa and Qa:
    - Aa most antigenic and produces peracute disease within 12–18 hours.
    - Qa not associated with clinical disease but produces weak positive during testing.
    - Aa and Qa negative are at a higher risk of producing a foal with NI.
- NI reported in 2% of Standardbreds and in less than 1% of Thoroughbreds.
- primary isoimmunisation by genetically incompatible erythrocytes occurs late in pregnancy:
  - antibody levels in mare's serum peaking at 9 days post-partum.
    - first foals are therefore not affected.
  - subsequent pregnancies, the secondary anamnestic response produces:
    - significant anti-erythrocyte antibodies concentrated in the colostrum.
    - puts the foal at risk.
  - tissue vaccines or transfusions may sensitise a mare to genetically incompatible erythrocytes.
  - antibodies concentrated in colostrum during the last weeks of pregnancy:
    - absorbed during the first 24 hours after birth.
    - antibodies pass into the foal's circulation.
    - attach to foal's erythrocyte surface, causing them to agglutinate:

- – removed by the monocyte–phagocytic system.
- – or haemolyse intravascularly in the presence of complement.
- disease only develops if:
  - ○ mare lacks certain red cell factors.
  - ○ mare is repeatedly exposed to that factor.
  - ○ mare conceives a foal that inherits that red cell factor from the sire.
  - ○ colostrum is produced containing anti-erythrocyte antibody.
  - ○ foal ingests and absorbs antibody, leading to immune-mediated destruction of its red cells.

## Clinical presentation

- foals appear healthy at birth.
- speed of onset and severity of clinical signs dependent on red cell antigen involved and the efficacy of passive immunity transfer:
  - ○ clinical signs usually develop between 12 and 48 hours post-partum.
  - ○ can be earlier and occasionally up to 96 hours.
- early clinical signs may be subtle and non-specific:
  - ○ lethargy, depression, recumbency, and reduced nursing.
  - ○ heart and respiratory rate increase when excited or restrained.
  - ○ icteric mucous membranes and sclera.
- severe clinical signs include:
  - ○ haemoglobinuria, rapid progression to collapse, coma, and/or death.

## Differential diagnosis

- other causes of haemolytic anaemia, including:
  - ○ drug administration, haemoparasites, toxin exposure, and sepsis.

## Diagnosis

- based on history, clinical signs, and laboratory results.
- Haematology:
  - ○ marked anaemia
    - ♦ RBC count $<4 \times 10^{12}/l$ ($<4 \times 10^6/\mu l$) haemoglobin $<70$ g/l ($<7$ g/dl).
    - ♦ PCV $<0.20$ l/l (20%) can be severe (PCV $<0.1$ l/l [<10%]).

- ○ thrombocytopaenia presents in some cases.
- plasma protein concentration remains within reference intervals.
- Biochemistry:
  - ○ unconjugated bilirubin concentration increased.
- haemoglobinaemia and haemoglobinuria if sufficient intravascular haemolysis exists.
- Direct Coombs test will be positive, although false negatives do occur.
- definitive diagnosis requires demonstration of alloantibodies in the mare's blood or colostrum directed against the foal's red cells:
  - ○ serum and red cells from the mare and foal are required for this test.
  - ○ offered by some commercial or specialist laboratories.

## Management

- treatment is dependent on the severity of clinical signs and degree of anaemia.
- mild cases:
  - ○ minimal handling and box rest (or very small nursery paddock turnout):
    - ♦ no stress or excitement.
  - ○ monitor RBC parameters twice daily for the first few days.
  - ○ slower falls in RBC numbers allow the foal to better adapt:
    - ♦ less obvious clinical signs and need for intensive treatment.
- severe cases:
  - ○ transfusion with donor red cells when PCV falls below 0.12 l/l (12%) and RBCs below $3.5 \times 10^{12}/l$ ($3.5 \times 10^6/\mu l$).
- early or rapid development of clinical signs such as depression, loss of suck, or signs of hypoxaemia, then earlier transfusion should be considered.
- volume of blood required can be calculated as below:

$$\text{Volume (ml)} = \frac{\text{body weight (kg)} \times 150 \times (\text{PCV desired} - \text{PCV observed})}{\text{PCV of donor}}$$

(150 is considered the blood volume in ml/kg for a 2-day old foal):
- 50 kg foal, it is usual to transfuse 2–4 l of blood.

- dam is usually considered the most suitable donor:
  - her red cells must be washed free of plasma containing antibody.
  - wash cells three times in saline and resuspend in isotonic saline.
- not possible to wash the cells, or the dam is not a suitable donor:
  - cross-matched gelding can be used as a source of whole blood.
- supportive therapy and nursing are an essential part of treatment for all cases:
  - intravenous electrolyte solutions encourage diuresis and prevent renal failure.

## Prognosis

- usually very good with appropriate treatment.
- foals with rapid haemolyis are more likely to require multiple transfusions:
  - associated with poorer prognosis due to likelihood of liver failure caused by iron overload.
  - lifespan of transfused red blood cells is usually between 4–7 days:
    - ◆ decreases with subsequent transfusions due to upregulation of foal's immune system.
    - ◆ Deferoxamine used as an iron chelator to prevent iron overload although efficacy is unclear.
- Kernicterus caused by marked hyperbilirubinaemia can be a severe complication:
  - clinical signs include abnormal muscle tone, opisthotonos and seizures, or sudden death.
  - effective treatment for kernicterus is difficult.

## Prevention

- 'High-risk' mares require preventive strategies for subsequent pregnancies:
  - possible to blood type the dam and sire to assess the risk of NI.
  - titrate a blood sample from the mare during the last 2–3 weeks of gestation for anti-erythrocyte antibodies:
    - ◆ do not take the sample too early as antibody levels rise late in gestation.
- foals from high-risk mares must be prevented from nursing:
  - given 500 ml of appropriate donor colostrum immediately after birth.
  - then muzzled prior to udder-seeking behaviour.
  - bottle-fed:
    - ◆ minimum of a further 500 ml of colostrum.
    - ◆ then milk replacer at the appropriate rate.
  - IgG levels checked at 18–24 hours post-partum to ensure they are adequate.
- mare's udder stripped frequently, and IgG concentration monitored until it falls:
  - stripped milk discarded.
- usually possible to allow the foal to return to the mare within 24 hours of foaling.

## PERINATAL / YOUNG FOAL CONDITIONS

# Neonatal sepsis

## Definition/overview

- life-threatening condition in the neonatal foal.
- untreated, sepsis can rapidly progress to septic shock and multiorgan dysfunction.
- early aggressive intervention increases the likelihood of a favourable outcome.

## Aetiology/pathophysiology

- Systemic inflammatory response syndrome (SIRS) describes the body's response to upregulation of the immune system following an inflammatory challenge:
  - sepsis is SIRS triggered by bacterial infection.
  - other factors also trigger SIRS including viral/fungal infection, tissue hypoxia and trauma.
- newborn foals are often exposed to a significant bacterial challenge shortly after birth.
- foal's immune response to bacterial pathogens is essential in preventing clinical sepsis:

- FTPI leaves foals at higher risk for sepsis when the bacterial challenge is high.
- most common route of bacterial entry into the foal's circulation is via the intestinal tract:
  ○ other routes include the respiratory tract, placenta *in utero*, and the umbilicus.
- Pathogen associated molecular patterns (PAMPs), present on bacteria, activate both the innate and adaptive immune systems:
  ○ cytokines are released that cause further upregulation of the immune system.
  ○ cascade initiated affects a wide variety of effector organs and tissue types:
    ♦ including coagulation system, endocrine system, and endothelium.
    ♦ responsible for most of the clinical signs seen with SIRS and sepsis.
- well-balanced immune response leads to removal of the bacterial pathogen from the circulation and a return to normal homeostasis:
  ○ excessive immune response without balance by the compensatory anti-inflammatory response syndrome (CARS) leads to uncontrolled SIRS (sepsis):
    ♦ rapidly leads to a syndrome of multiple organ dysfunction and death.
- most common bacteria implicated in neonatal sepsis are all environmental pathogens:
  ○ include *E. coli*, *Enterobacter* spp., *Staphylococcus* spp., *Streptococcus* spp., *Actinobacillus* spp., *Klebsiella* spp., *Pasteurella* spp., *Salmonella* spp., *Clostridium* spp., and *Enterococcus* spp.).
  ○ many cases are caused by mixed infections.

## Clinical presentation

- clinical signs are variable:
  ○ range from subtle, non-specific, and insidious in onset to acute and fulminant.
- less frequent feeding to complete anorexia.
- lethargy and increasing periods of recumbency.
- altered mental status such as depression and seizures.
- increased heart and respiratory rates.
- increased respiratory effort and distress.
- raised or subnormal temperature (normal does not rule out sepsis).
- congestion and petechiation of mucous membranes.
- coronary band hyperaemia (Fig. 3.1).
- diarrhoea
- colic and abdominal distension.
- uveitis with ocular pain (Fig. 3.2).
- hypovolaemia.
- bone or joint infections with joint effusion and lameness.
- umbilical infections and patent urachus.
- cardiovascular collapse, coma, and death.

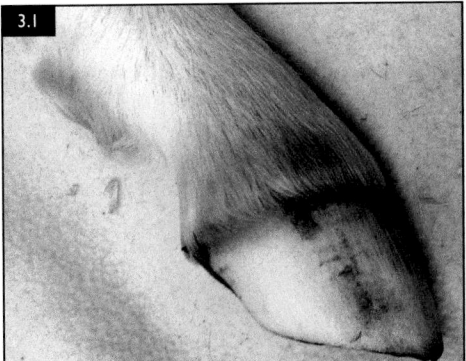

**FIG. 3.1** Distinct hyperaemia of the coronary band can be an indication of sepsis.

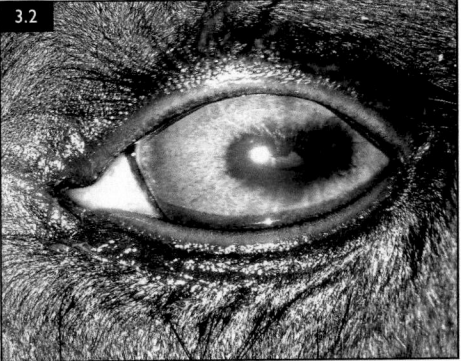

**FIG. 3.2** A septic uveitis of the left eye of a foal with septicaemia. Note the marked miosis of the pupil.

## Differential diagnosis

- Neonatal maladjustment syndrome
- metabolic disturbances.

## Diagnosis

- based on history and a thorough clinical examination combined with clinical pathology:
  - **sepsis should always be considered until proven otherwise.**
  - **if suspected, initiate treatment immediately – positively affects outcome.**
- positive blood culture is the gold standard for diagnosis:
  - technique is time consuming and false negatives are common.
  - samples from other sources can also be cultured including:
    - joints/synovial structures, tracheal washes, urine, faeces and CSF.
- **sepsis scoring system** can be used to help increase the accuracy of diagnosis:
  - collection of several clinical and haematological/biochemical assay results.
  - each are allocated scores, the sum of which predicts the likelihood of sepsis.
- **laboratory findings:**
  - leucopaenia and neutropaenia are common, often with a left shift.
  - increased acute phase proteins:
    - serum amyloid A (SAA) (rises over 12–24 hours) – good early indicator.
    - fibrinogen (rises more slowly over 48 hours).
  - IgG concentration checked to determine whether FTPI is present.
  - biochemical evaluation useful to evaluate for evidence of organ dysfunction:
    - liver enzymes are commonly mildly increased in foals with sepsis.
  - blood glucose, lactate, and electrolytes should be frequently monitored:
    - hypoglycaemia is common in the early stages of sepsis:
    - frequently followed by hyperglycaemia due to endocrine dysfunction.
    - hyperlactataemia is common due to poor tissue perfusion.

## Management

### Antibiotics

- broad-spectrum bactericidal antibiotics with good penetration of affected tissues.
- provided renal function is not significantly compromised:
  - penicillin/ampicillin and an aminoglycoside (gentamicin/amikacin) are a commonly used first choice.
  - dose rates appropriate for neonatal foals should be used (Table 3.2).
  - adjust antimicrobial choice based on results of culture/sensitivity.

### Cardiovascular support

- tissue perfusion and adequate oxygen delivery to the tissues is important.
- monitoring hydration, blood pressure, mental status, and blood lactate concentrations provides useful information.
- resuscitation fluid boluses (10–20 ml/ kg of balanced electrolyte solution at a higher flow rate [i.e. 1 litre/50 kg foal over 20 minutes]) provide temporary volume expansion:
  - monitor response to therapy – further boluses (maximum 4) may be necessary:
    - **foals are more susceptible to volume overload than adults.**
- plasma may also be useful during resuscitation.
- failure to respond to appropriate fluid therapy indicates the need to use ionotropes and vasopressors to support blood pressure.

### Respiratory support

- prolonged periods of recumbency, poor tissue perfusion, and weakness contribute to ventilation/perfusion mismatch and atelectasis.
- most sick collapsed foals benefit from humidified intranasal oxygen therapy.
- maintain the foal in sternal recumbency.
- coupage encourages drainage of secretions.

### Nutrition

- enteral feeding is poorly tolerated in many cases:
  - parenteral nutrition should be considered at an early stage.

- mare's milk is most appropriate for enteral feeding if this is tolerated:
  - if suck reflex is weak, an indwelling feeding tube allows low-volume frequent feeding.
  - hourly for the first 2–3 days, decreasing to every 2 hours by 3–4 days post-partum.
  - volumes should be small at first (100–200 ml/feed) and gradually built up if tolerated.
  - healthy foal consumes about 20–23% BWT per day in milk to fulfil requirements:
    - sick foals require less and about 10% of BWT per day should be worked towards over several days.
    - maximum of 500 ml/per feeding is recommended.
    - **important to closely monitor foal for ileus as volume of milk fed increases.**

## Other drugs

- use of anticoagulants such as low molecular weight heparin may be beneficial.
- Polymixin B can also be used as an anti-endotoxic drug (potential for nephrotoxicity).
- reduced doses of flunixin meglumine (0.25 mg/kg i/v q8h) have been used in the treatment of septic shock, but the evidence for efficacy remains controversial:
  - **potential for causing serious side effects, (nephrotoxicity, and gastroduodenal ulceration), especially in hypovolaemic and collapsed foals.**
- **High standards of hygiene and nursing care** are an important part of therapy:
  - includes the careful regulation of the environmental temperature.

## Prognosis

- mortality rates can be high:
  - survival is variable depending on factors such as:
    - speed of referral to a critical-care centre.
    - level of care and the financial resources available.
- survival rates in the field have not been documented.

# Prematurity/dysmaturity

## Definition/overview

- normal gestation in the horse is approximately 335–345 days:
  - variation between breeds and individuals.
- gestational age has been used to define this condition in foals:
  - <320 days as premature.
  - normal/prolonged gestation but characteristics of prematurity, as dysmature.
- wide normal variation in gestational length (and potential errors in breeding records):
  - each foal evaluated individually.
  - greater reliance on clinical signs than gestational age.
- most foals born prior to 320 days require some veterinary intervention:
  - 280 days is considered the cut-off for survival.
- degree of maturity of the various body systems may be asynchronous.

## Aetiology/pathophysiology

- final cortisol surge in the foal, which is responsible for maturation of all vital body systems, occurs in the 48 hours prior to parturition.
- foals born before this cortisol surge (e.g. after inappropriate induction of parturition) are often not viable and succumb to multiple organ dysfunction.
- foals that have undergone chronic *in utero* stress (e.g. placentitis) have often received more maturational signals and can be viable at a much younger gestational age.
- causes of prematurity include:
  - placental insufficiency
  - placentitis
  - twinning.
  - maternal disease or early induction of parturition.

## Clinical presentation

- foals present with some or all the following physical characteristics (Fig. 3.3):
  - small and underweight
  - poor body condition
  - slow to stand and suck.
  - silky mole-like coat

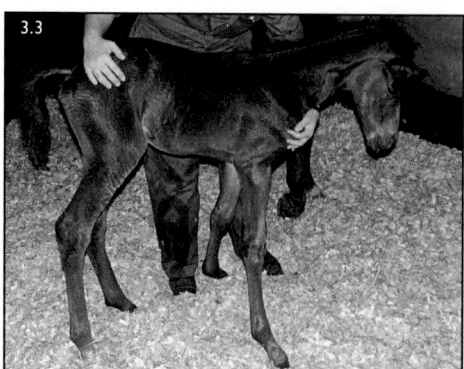

**FIG. 3.3** A foal showing some of the classic signs of prematurity including a domed forehead, low body weight, silky coat and floppy ears.

- poor thermoregulation.
- domed forehead
- floppy ears and discoloured tongue.
- flexor tendon laxity and weak musculature.
- pale mucous membranes
- weakness and occasionally incoordination.
- further investigations may reveal:
  - incomplete ossification of cuboidal bones and joint instability.
  - poor tolerance to enteral feeding, with colic and abdominal distension.
  - poor respiratory function with slow rate and/or abnormal respiratory pattern.
  - altered mentation
  - increased susceptibility to infection.
  - dehydration.
  - leucopaenia and neutropaenia
  - neutrophil:lymphocyte ratio <2:1.
  - hypoxia and hypercapnia   ○ acidosis.

## Differential diagnosis

- sepsis  • perinatal asphyxia syndrome.

## Diagnosis

- based on history of the pregnancy, physical appearance, and haematological parameters.
- **ACTH stimulation test:**
  - mature foals show an increase in total WBC count and neutrophil:lymphocyte ratio (>2:1) in response to short-acting ACTH (Synacthen).

- cortisol concentration should double in mature foals.
- radiographs of the lungs and limbs are helpful (Fig. 3.4):
  - assess severity and type of respiratory abnormalities (e.g. atelectasis).
  - cuboidal bones within carpus/ tarsus to assess degree of incomplete ossification.

## Management

- intensive and supportive treatment with an enormous input of nursing care is required:
  - **owners of such foals need to be made aware of this at the outset.**

### Prenatal corticosteroids

- opportunity for their use in the mare is limited and the most appropriate dose and drug have not been established.
- large doses of dexamethasone have been used in some mares to hasten the onset of parturition and foetal maturation.

### Endocrine support

- use of corticosteroids in the newborn remains controversial:
  - optimum dose and drug have not been established.
  - hydrocortisone can be given at physiological doses (1.3mg/kg/day divided into four-hourly doses).
  - other suggested regimens include:
    - single 6 mg dose of betamethasone for a 50 kg foal.
    - dexamethasone at a dose of 0.02 mg/kg once daily.

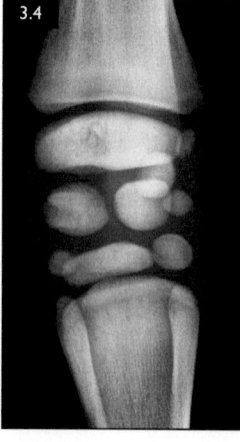

**FIG. 3.4** Dorsopalmar radiograph of incomplete ossification of the carpal bones of a premature foal. Note the rounded outline of the osseous centres of the carpal bones and the apparent large space in between them.

## Nutritional support

- premature foals may not tolerate enteral feeding – used cautiously and monitored.
  - colic, ileus and necrotising enteritis can result.
- premature foals are often in poor body condition and prone to hypoglycaemia:
  - use of intravenous glucose and/ or parenteral nutrition may be appropriate.
  - careful monitoring of blood glucose is important.
  - continuous intravenous infusion of insulin may be required.
- combination of enteral feeding and parenteral nutrition:
  - required in many cases to maintain calorie intake and encourage GI maturation.
  - small trophic feeds of 5–10 ml/hour of milk.
  - oral glutamine (10 g/day in divided doses) may help enterocyte repair/ function.

## Metabolic and cardiovascular support

- intravenous fluids are frequently required but volume overload can occur quickly:
  - these foals are very intolerant of excess quantities of sodium.
  - some foals require greater support with the use of inotropes or vasopressors.
  - homeostatic mechanisms may be poorly established:
    - electrolyte concentrations should be monitored regularly.
- foals should be warmed, or the environmental temperature increased as required.
- premature foals can suffer 'second day syndrome':
  - after 24 hours of improvement, they slide irreversibly into hypotension, hypoxia, septic shock, and multiorgan dysfunction syndrome.

## Immunological support

- premature foals often have FTPI and are particularly susceptible to infection:
  - broad-spectrum bactericidal antibiotics and intravenous hyperimmune plasma are helpful.

## Respiratory support

- humidified intranasal oxygen may be required.

- maintain the foal in the sternal position or regularly turned from side to side.
- if it can stand, should be assisted in doing so.
- arterial blood gas analysis is helpful in monitoring respiratory function:
  - failing respiratory function is a poor prognostic sign.
  - mechanical ventilation may be required in the more severely affected foals.

## Musculoskeletal support

- degree of cuboidal bone ossification is an indicator of maturity.
- foals with poorly ossified carpal and tarsal bones can develop severe crushing of the cuboidal bones and severe limb deformity.
- restricted exercise, maintaining steady growth, and good farriery are required.
- severe cases:
  - limb splint bandages, casts, or custom-made limb supports.
  - these individuals carry a poor prognosis for future athletic performance.

## Prognosis

- survival after 300 days gestation is influenced by whether the pathology resulting in the premature delivery was acute or chronic:
  - chronic placental pathology cases have a greater chance for survival.
- sepsis and metabolic dysfunction are common complications:
  - often leads to rapid deterioration and death 48–72 hours after birth.
- **important to consider the future use of the horse:**
  - musculoskeletal problems may preclude an athletic future.
- no individual parameter is the key to the prognosis and a whole range should be considered before deciding on the best course of treatment.

# Meconium aspiration syndrome

## Definition/overview

- usually associated with maternal or foetal stress and is uncommon.

- severity can vary from mild respiratory compromise to fatal respiratory distress.

## Aetiology/pathophysiology

- foetus can pass meconium during periods of stress.
- either the foetus gasps *in utero* or the fluids are not cleared from the airways prior to the first breath (Fig. 3.5):
  - meconium-contaminated amniotic fluid is aspirated into the lungs.
  - produces a chemical pneumonitis.

## Clinical presentation

- at birth:
  - foal and nostrils are covered in liquid brown meconium.
  - amniotic fluid is stained yellow/brown.
- respiratory distress occurs within the first few days of life:
  - often an increased respiratory effort.
- some foals may exhibit signs of neonatal maladjustment syndrome.
- degree of hypoxaemia and respiratory acidosis is dependent on the quantity of fluid aspirated and the degree of lung injury.

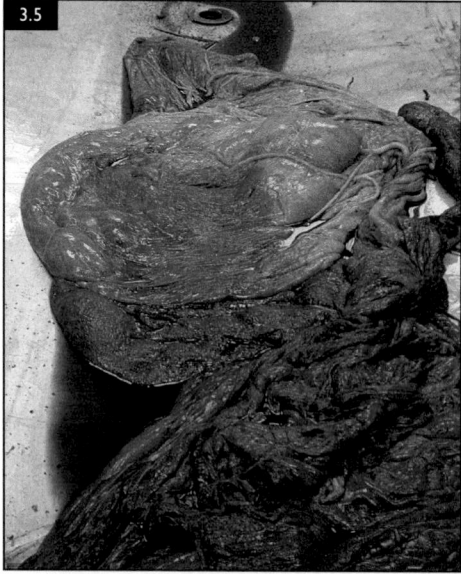

**FIG. 3.5** A meconium-stained placenta from a mare whose foal suffered meconium inhalation.

## Diagnosis

- history and clinical signs provide a presumptive diagnosis.
- visualisation of brown-stained fluid in the trachea with endoscopy.
- granular appearance of caudoventral lungs on radiographs.
- confirmation is based on histological examination of the lungs.

## Management

- early diagnosis allows attempts to be made to suction as much meconium-contaminated fluid from the airways as possible – **care not to induce further lung damage.**
- humidified intranasal oxygen can reduce the work of breathing.
- corticosteroid use remains controversial.
- high levels of supportive care may be required:
  - broad-spectrum bactericidal antibiotics should be given.
- mechanical ventilation can be used in severe cases.

# Neonatal maladjustment syndrome (NMS)

## Definition/overview

- affects foals less than 3 days of age.
- most common clinical signs are neurological but many body systems can be affected:
  - hence foals with this condition exhibit a wide range and severity of clinical signs.

## Aetiology/pathophysiology

- two recognised scenarios that give rise to this clinical condition:
  - foals with a hypoxic–ischaemic injury that occurs around the time of foaling (e.g. difficult delivery or dystocia) – **Perinatal Asphyxia Syndrome (PAS):**
    - ♦ these foals often have the most severe clinical signs and a worse prognosis.
    - ♦ hypoxic injury to a variety of organ systems is seen at post mortem:
      - – most commonly the CNS, kidney, and gastrointestinal system.

- foals that have no history or suspicion of birth hypoxia – **NMS**:
  - fail to fully adapt to extra-uterine life.
  - show persistently high or abnormal progestogen concentrations after birth.
  - exact cause is unknown but factors such as systemic inflammation, hypoxia, abnormal delivery, and placental abnormalities are thought to play a role.
  - in some foals, it has been observed that foaling is unusually rapid:
    - raised the possibility that the physical pressure of the birth canal may be an important stimulus for normal maturational pathways.
  - progestogen concentrations fall in line with recovery in these NMS foals:
    - usually occurs within 2–5 days of supportive care.
  - post-mortem of non-survivors:
    - usually, unremarkable aside from changes related to secondary sepsis and perfusion abnormalities.

## Clinical presentation

- foals may be normal at birth and then develop clinical signs within 12–48 hours:
  - these foals tend to be the less severely affected.
  - those foals that show signs immediately after birth tend to have more severe disease and a poorer prognosis.
- neurological signs are often the most common:
  - range from a foal that appears slightly 'slow' to a foal with marked obtundation or uncontrollable seizures.
  - common signs include:
    - loss of suck reflex and poor teat searching.
    - persistent chewing movements
    - tongue protrusion (Fig. 3.6).
    - aimless wandering, altered mentation and abnormal head carriage.
    - reduced interaction with the mare or the environment.

- hyperaesthesia, weakness or rapid exhaustion.
- inability to stand or stay standing.
  - more severe signs include:
    - central blindness with anisocoria, opisthotonus or hypotonia, seizures and coma.
- other body systems commonly affected include:
- **Respiratory system:**
  - periods of apnoea, abnormal breathing patterns, or barking vocalisation (rare), shallow tachypnoea and/or dyspnoea can be seen.
  - some of these signs relate to abnormal central control of breathing rather than primary or secondary respiratory disease.
- **Gastrointestinal system:**
  - mild colic to severe ileus:
    - reduced intestinal transit time and faecal output.
    - other foals become completely intolerant of enteral feeding.
- **Urinary system:**
  - renal signs range from mild oliguria to complete anuria.
  - latter is difficult to manage and often associated with high mortality.
- **Endocrine and metabolic dysfunction** is also common.

## Differential diagnosis

- numerous differential diagnoses including neonatal sepsis, septic meningitis,

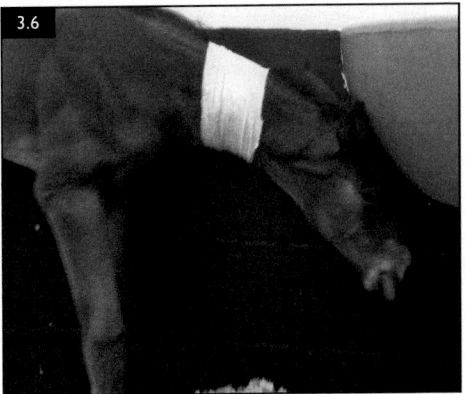

**FIG. 3.6** A foal showing classic early signs of NMS with no affinity for the mare and a protruding tongue.

severe metabolic disturbances leading to electrolyte imbalances and renal failure, cranial trauma, and hydrocephalus.

## Diagnosis

- history, clinical signs and ruling out other conditions.
- clinicopathological abnormalities reflect the presence and severity of complications.

## Management

- majority of foals with NMS survive if given early and appropriate supportive care:
  - once secondary complications develop, such as sepsis, the survival rate decreases.
- main aims of treatment are:
  - maintain hydration
  - provide nutrition    - prevent sepsis.
  - encourage maternal bonding
  - control seizures or other neurological signs.

### Cardiovascular support

- important to maintain adequate tissue perfusion:
  - involves appropriate fluid therapy to restore and maintain circulating volume.
  - inotropes and pressors may be necessary in more severe cases to maintain blood pressure and tissue perfusion.

### Nutritional support

- mild NMS foals will often tolerate enteral feeding:
  - many have a poor suck reflex and should not be fed from a bottle.
  - use of a small indwelling nasogastric tube can allow small, frequent feedings.
  - monitor very closely as some foals will not tolerate large volumes of milk.
    - 10% of bodyweight is a good initial target.
    - provides maintenance fluid and energy requirements.
- more severe cases will not tolerate enteral feeding:
  - parenteral nutrition with small trophic feeds is required.

### Respiratory support

- recumbent or severely affected foals will benefit from respiratory support:
  - humidified intranasal oxygen at 2–15 litres/minute.
  - sternal positioning and encouraging mobility can significantly improve lung function.
  - arterial blood gas analysis useful to monitor respiratory function.
  - foals with abnormal central control of respiration will often develop excessive hypercapnia (without hypoxemia):
    - use of respiratory stimulants such as oral caffeine (10 mg/kg p/o followed by 2.5 mg/kg p/o q24h) or doxapram CRI can reduce hypoventilation.

### Control of seizures

- diazepam or midazolam (0.1–0.2 mg/kg i/v) for control of seizures in the short term:
  - accumulate in tissues and are not suitable for longer term use at this dose.
  - midazolam by CRI (0.02–0.06 mg/kg/hour) is excellent for more prolonged treatment (6–72 hours+).
  - phenobarbitone (2–10 mg/kg by slow i/v infusion up to q8h) reduces CNS excitability and can be used for control of frequent, more severe seizures:
    - disadvantage is its prolonged effect and significant respiratory depression.
  - phenytoin is an alternative anti-seizure drug.

### Other treatments

- antioxidants such as Vitamin E, C and thiamine suggested to reduce oxidative damage.
- magnesium sulphate to prevent cellular death.
- DMSO, mannitol and dexamethasone to reduce cerebral oedema.
- all have minimal evidence to support any of their use.
- broad-spectrum antimicrobial therapy should be considered in these cases.
- FTPI should be addressed if present.
- **'squeeze-induced somnolence' or the 'foal squeeze' technique** (Fig. 3.7):

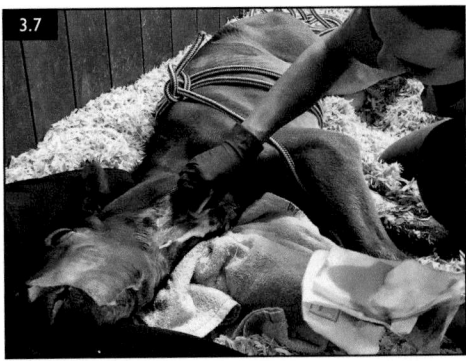

**FIG. 3.7** The use of squeeze-induced somnolence to treat a foal with clinical signs of NMS.

- developed recently to try to recreate the physical pressure of the birth process.
- may lead to a dramatic and rapid improvement in the clinical signs in some foals.

## Nursing

- high standards of nursing care and hygiene are required.
- quiet warm environment, if possible close to the mare, is essential.
- manual restraint on a padded bed to prevent self-inflicted trauma is important.
- more severely affected cases, need encouragement and considerable patience, time, and skill to:
  - rise, move and regain the suck reflex.
  - interact with the environment, and suck from the mare.

## Prognosis

- survival rates range widely but can be as high as 80–90%.
- foals that experience significant severe hypoxic injury is generally poorer.
- uncontrollable seizures, anuric renal failure and severe cardiovascular dysfunction are generally associated with a poor prognosis.

## Atresia coli/recti/ani

### Definition/overview

- rare congenital abnormalities:
- part of the hindgut fails to develop *in utero*.

- obstruction of the passage of faeces leads to clinical signs of colic, usually in the first 1–2 days of life.
- condition is fatal in most cases.

## Aetiology/pathophysiology

- developmental abnormalities of one or more segments of the GI tract:
  - atresia coli, and to a lesser extent atresia ani, are the most common.
  - former characterised by membranous occlusion of the lumen, remnants of gut connecting two blind ends (cord atresia), or the presence of blind ends with no connection.
  - atresia ani – absence of an anus and variable parts of the rectum.
- concurrent urinary tract defects are rare.

## Clinical presentation

- foals are normal at birth and usually stand and suck normally.
- usually present within 4–24 hours with a progressive, moderate to severe colic:
  - physical obstruction of the passage of gas and faeces.
  - more caudal the defect, the slower the onset of signs.
  - atresia ani, no anus is visible, or it appears grossly abnormal.
  - obstructions elsewhere, digital examination per rectum reveals no faeces present or palpable.
  - abdominal distension, anorexia, and colic progress over a period of hours.

## Differential diagnosis

- meconium retention
- abdominal crisis (e.g. small intestinal volvulus).
- ruptured bladder.
- ileocolonic aganglionosis (lethal white syndrome)
- ileus.

## Diagnosis

- based on clinical signs, physical and digital examination (diagnostic for atresia ani).
- abdominal ultrasonography may confirm colonic obstruction or possible narrowing.
- plain radiographs of the abdomen may demonstrate faecal retention and/or

gaseous distension proximal to the site of the intestinal occlusion.
- contrast radiography with a barium enema useful for confirming the location of the atresia.
- proctoscopy or colonoscopy will usually confirm the lesion.
- exploratory laparotomy is occasionally required to confirm diagnosis of more proximal lesions and to determine what (if any) treatment options are available.

## Management

- true atresia ani, surgical reconstruction of the anus may be possible.
- colostomy and anastomosis techniques have been attempted but results have been poor.
- most cases are not amenable to surgery and the prognosis is hopeless.

# Meconium retention

## Definition/overview

- Meconium consists of digested amniotic fluid and cell debris accumulated during foetal life:
  - usually passed within 12 hours of birth.
  - pelleted or occasionally tarry, dark brownish/green or black faecal material.
  - precedes normal milk faeces.

## Aetiology/pathophysiology

- retention most commonly occurs in the rectum but also more proximally in the colon.
- seen in foals suffering from NMS.
- why the condition affects some foals is not clear.

## Clinical presentation

- usually present with mild to severe colic from 6–36 hours post-partum:
  - persistent unproductive straining to pass faeces, with tail swishing or lifting.
    - squatting, and crouching (rounded back).
  - foals develop progressive gas abdominal distension.
  - intermittently, or completely, off suck.

- passage of some meconium does not rule out meconium retention.

## Differential diagnosis

- congenital deformities of the GI tract present similarly, whereas other abdominal conditions are usually easily differentiated.

## Diagnosis

- clinical examination and history are important.
- careful digital rectal examination may reveal pellets of meconium.
- ultrasonography of the abdomen is very useful to visualise the meconium:
  - very echogenic (almost sparkly) round balls within the bowel (Fig. 3.8)
  - site of retention.
- barium enema radiographs useful to confirm the diagnosis in refractory cases.

## Management

- good colostrum intake has a laxative effect and may help to prevent this condition.
- primary treatment involves the use of enemas:
  - buffered phosphate enemas are effective in mild cases.
    - repeated use may result in hyperphosphataemia and an inflamed rectal mucosa.
  - soap and water enemas can also be used.
- use of mineral oil by nasogastric tube should be avoided.

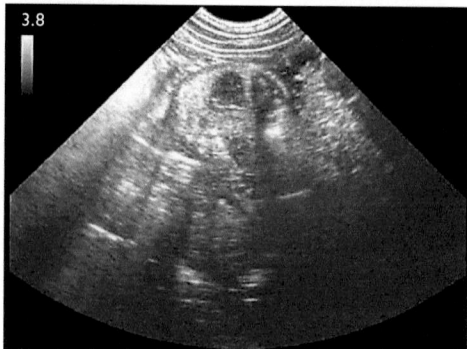

FIG. 3.8 Image of an ultrasonographic examination of the cranial abdomen confirming the presence of meconium within the bowel.

- refractory cases:
  - use of 4% buffered acetylcysteine enemas is usually successful:
    - 100–200 ml of warmed solution administered slowly via a 30 French gauge Foley catheter with the balloon inflated with 30 ml of air.
    - powerful mucolytic that helps break down the meconium.
    - retained in the rectum for 30–40 minutes if possible.
    - may take several hours to work and occasionally may need repeating.

- use of analgesics and ensuring adequate hydration is essential.
- good nursing care, particularly protection against self-inflicted trauma, is important.
- use of fingers or instruments to remove pellets from the rectum is contraindicated.

## Prognosis

- usually good to excellent with correct management and without complications.
- surgery is almost never indicated.

## CONDITIONS AFFECTING BOTH YOUNG AND OLDER FOALS

## Gastroduodenal ulceration disease (GDUD)

### Definition/overview

- affects foals of all ages and its prevalence varies between 25–57%.
- clinical signs are typically more severe than in mature horses.

### Aetiology/pathophysiology

- gastric environment in the young foal is highly acidic by 1 week of age:
  - disruption of mucosal protective mechanisms can result in ulceration of the stomach or duodenum.
- risk factors include:
  - physiological stress
  - hypoxia (neonatal foals).
  - delayed gastric emptying.
  - administration of NSAIDs.
  - prolonged feeding intervals.
  - prolonged recumbency.
- associated with gastrointestinal disease or as a complication of diarrhoea:
  - particularly rotavirus diarrhoea.
- lesions are most common in the squamous mucosa at the margo plicatus.
- rupture of a subclinical gastric ulcer leading to septic peritonitis and sudden death is rare, but usually fatal, and often associated with other severe neonatal disease.

### Clinical presentation

- number of different presentations in foals:

1. **Subclinical small squamous lesions** (usually at the margo plicatus):
   - incidental finding if gastroscopy is performed.
   - epithelial desquamation also identified as a normal part of mucosal maturation in healthy foals up to 3 months of age.
2. **Large or significant gastroscopic lesions:**
   - variable clinical signs include anorexia and reduced nursing, bruxism, ptyalism (Fig. 3.9), dull demeanour, colic, poor growth, rough hair coat, or diarrhoea.
3. **Perforated gastric ulceration** leading to peracute septic peritonitis, distended abdomen, colic, and sudden or unexpected death in some cases:
   - premonitory GDUD clinical signs may or may not be present.
   - hospitalised foals with concurrent disease most at risk.
4. **Pyloric or duodenal ulceration** (most common in 2–5-month-old foals).
   - severe inflammation, fibrosis and resultant stenosis of the gastric outflow tract and duodenum leading to delayed gastric emptying:
     - colic, ptyalism, eructation, anorexia, failure to thrive and dull demeanour.

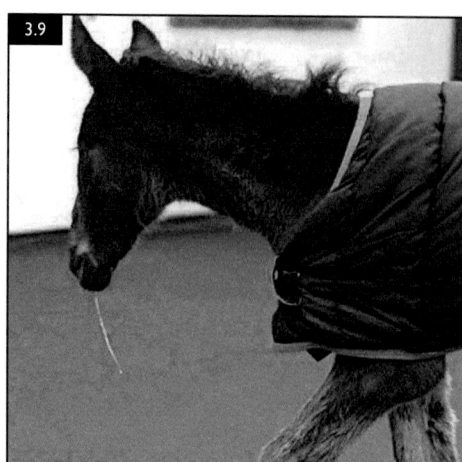

FIG. 3.9 A foal with gastric ulceration showing hypersalivation.

- reflux easily obtained during nasogastric intubation.
  - some cases, arises spontaneously.
- gastroscopy often also identifies reflux oesophagitis and oesophageal ulceration.

## Differential diagnosis

- other causes of colic such as enteritis, septicaemia, and peritonitis may produce similar signs.

## Diagnosis

- history and clinical signs are often strongly suggestive of GDUD.
- confirmed with gastroscopy:
  - older foals with clinically complex gastric disease and/or a poor response to treatment.
  - traumatic and invasive for young foals and so rarely performed.
- positive contrast radiography (barium administered by stomach tube) can evaluate gastric transit times but may be difficult to interpret.
- GGT and serum alkaline phosphatase blood biochemistry will be raised if strictures occur close to the bile duct.
- abdominocentesis will reveal evidence of septic peritonitis associated with a perforated lesion:
  - lesion often only definitively confirmed during exploration of the abdomen at surgery or post-mortem.

## Management

- mainstay of treatment is acid suppression:
  - need for treatment should consider disease severity and overall clinical picture.
  - continue treatment until clinical signs or lesions have completely resolved:
  - may be several weeks.
  - response to treatment is disappointing for pyloric and/or duodenal stenosis.
- underlying/other disease should be treated.
- foals with gastroduodenal stenosis may require repeated nasogastric decompression and in severe cases may only recover with gastrojejunostomy bypass surgery.
- various drugs have been used to treat gastroduodenal ulcers:
  - **Proton pump inhibitors:**
    - Omeprazole (4 mg/kg p/o q24h).
    - multiple licensed omeprazole formulations are available.
    - treatment of choice and clinical signs will often improve within days of medication.
    - most cases of squamous disease will resolve in 2-4 weeks:
      - glandular or duodenal gastric disease often take longer.
  - **H2-receptor antagonists:**
    - must be given at adequate dose rates and frequencies to be effective.
    - Ranitidine (6.6 mg/kg p/o q8h).
      - intravenous formulations 1–2 mg/kg q8h:
      - may be necessary when the condition of the foal precludes oral use.
    - Cimetidine (16–20 mg/kg p/o q6h) is of questionable efficacy.
  - **Sucralfate:**
    - often used alongside omeprazole at 20 mg/kg p/o q6h.
    - mechanisms of action:
      - adherence to ulcerated mucosa.
      - stimulation of mucous secretion.
      - prostaglandin E synthesis.
      - enhanced blood mucosal blood flow.
    - important for glandular lesions where presumed aetiology is loss

of protective factors within the glandular mucosa.
- ○ Others:
  - ◆ depending on age of the foal and the severity of clinical signs:
    - – supportive care and judicious use of analgesics (opioids).

## Prevention

- allow normal feeding habits with the mare and foal.
- careful use of NSAIDs in the foal.
- prophylactic use of antiulcer medication in the foal remains controversial:
  - ○ reported associated with increased risk of developing diarrhoea in neonatal intensive care cases.
  - ○ considered in foals subjected to stressors such as surgery, other drug therapy, or other diseases.

## Prognosis

- good with early diagnosis and effective treatment.
- poorer if adhesions or pyloric/duodenal stenosis are present.
- perforated ulcers have a grave prognosis.

## Umbilical infections (ompalitis, omphaloarteritis, omphalophlebitis)

### Definition/overview

- infection of the external umbilical remnants or the internal umbilical vessels is quite common in the foal.

### Aetiology/pathophysiology

- infection occurs due to either:
  - ○ bacterial contamination of the external umbilical remnants during the peripartum period or:
  - ○ bacteraemia leading to deposition of bacteria in thrombosed vessels.
- infection forms either:
  - ○ focal external abscess (with/without surrounding body wall cellulitis).
  - ○ extends into the thrombi and infects the intra-abdominal vessels and/or urachus.
- many cases involve both internal and external umbilical remnants.

- infection of the umbilical vein may result in hepatic abscess formation.
- generalised septicaemia or localised bacterial infection (e.g. pneumonia and septic arthritis/osteomyelitis) is rare.
- bacteria most involved are those seen in neonatal septicaemia (i.e. Enterobacteriaceae, *Streptococcus* spp., and *Staphylococcus* spp.).

### Clinical presentation

- variable, and a thorough examination is important.
- enlargement of the external umbilicus (+/- draining purulent material):
  - ○ hot and painful to palpate.
  - ○ indicates focal abscess formation.
  - ○ may extend into the body wall if deeper structures also involved.
- infection of the internal umbilical remnants cannot be detected clinically:
  - ○ identified during diagnostic evaluation for a septic focus.
  - ○ more than one affected umbilical vessel is common in the neonate.
  - ○ urachal involvement is frequent.
- some foals may be pyrexic and inappetent.
- signs of SIRS, sepsis, or the sequelae (i.e. septic arthritis, uveitis) may be the first presenting clinical abnormalities.

### Differential diagnosis

- Umbilical rupture, haemorrhage or hernia
- patent urachus
- ventral oedema or cellulitis.

### Diagnosis

- clinical signs and umbilical remnant palpation.
- culture and sensitivity testing of draining purulent discharge.
- external remnant and transabdominal ultrasonography (Figs. 3.10–3.12):
  - ○ indicated for all foals with evidence of localised or systemic infection.
  - ○ internal vessels visible for about 4 weeks after birth.
  - ○ normal umbilical vein runs cranially to the liver and <10 mm in diameter.
  - ○ umbilical arteries run caudally to the bladder:
    - ◆ normally <12 mm in diameter.
  - ○ infected vessels are enlarged:

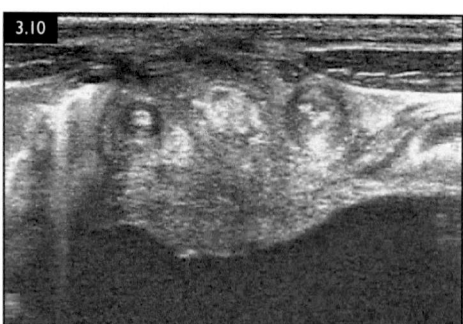

FIG. 3.10 Transabdominal ultrasonogram of the ventral abdomen of a foal showing a cross-section of the umbilical region with marked omphalitis. Note the two normal umbilical arteries either side of a central enlarged and inflamed umbilicus. (Photo courtesy Massimo Magri)

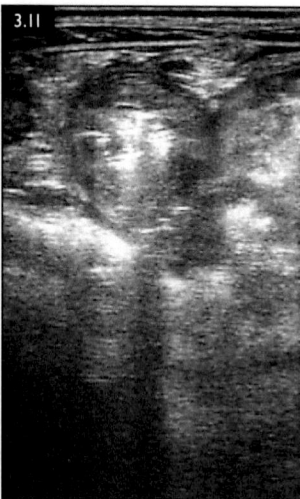

FIG. 3.11 Transabdominal ultrasonogram of the ventral abdomen of a foal with omphalophlebitis, or infection of the umbilical vein, with an enlarged single vein filled by hyperechoic material. (Photo courtesy Massimo Magri)

  ◆ contain hypoechoic, echogenic or anechoic material.
  ◆ may have thickened walls.
- severity of disease, the immune status of the foal, and presence of SIRS/sepsis:
  ○ determined by evaluating bloodwork for:
    ◆ leukocytosis or leukopaenia.
    ◆ increased acute phase proteins (SAA and fibrinogen).
    ◆ IgG levels.
  ○ some cases of localised external infection show minimal laboratory abnormalities.

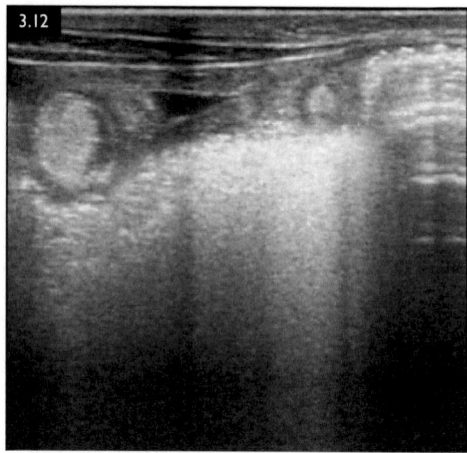

FIG. 3.12 Transabdominal ultrasonogram of the ventral abdomen of a foal with unilateral omphaloarteritis, or infection of one of the umbilical arteries, with enlargement of the artery and increased luminal hyperechoic material. (Photo courtesy Massimo Magri)

## Management

- external focal infection without signs of systemic involvement:
  ○ encourage abscess drainage (external).
  ○ up to 14–21 days of broad-spectrum bactericidal antibiotics.
    ◆ ideally based on sensitivity testing.
  ○ careful monitoring with repeat transabdominal ultrasonography and bloodwork.
- significant internal umbilical remnant infection, and/or poor response to medical therapy, or overwhelming signs of systemic involvement:
  ○ surgical resection of internal/external umbilical remnants is likely to be required.
- important preventive measures:
  ○ clean foaling environment.
  ○ correct care of the umbilicus after birth.
  ○ early ingestion of high-quality colostrum.
  ○ minimise foal contact with contamination sources (wash teats/udder pre-foaling).

## Persistent or patent urachus

### Definition/overview

- Patent urachus:

- urachus is open and urine leaks from the external umbilical remnants.
- result of a persistent urachus (failure to seal after birth) or may be acquired (reopening and resumed patency after an initial urachal closure).

## Aetiology/pathophysiology

- foetal urachus drains urine from the bladder into the allantoic cavity:
  - normally closes at parturition and urine flow ceases within 24 hours.
- possible causes of urachal reopening or failure to close include:
  - umbilical disorders such as torsion, increased length, early severance, or ligation rather than natural rupture.
  - excessive foal straining (i.e. meconium retention or ruptured bladder).
  - excessive or improper lifting of the foal's abdomen.
  - localised urachal/umbilical infections or systemic infections.
  - prolonged recumbency and reduced movement or stall confinement.
- patent urachus is a common complication in foals requiring intensive care.
- also occurs when irritants (e.g. silver nitrate) are applied to the external umbilical remnant, leading to tissue necrosis and vessel exposure.

## Clinical presentation

- urine observed leaking from the umbilical remnants continually or as the foal urinates:
  - foal may posture/strain frequently to urinate due to irritation or local infection of urachus.
- umbilical area is continually wet:
  - may be urine scalding and localised dermatitis.
- occasionally, the external umbilical remnants are necrotic (Fig. 3.13).

## Differential diagnosis

- other umbilical disorders
- ventral abdominal oedema
- localised cellulitis.

## Diagnosis

- clinical signs but bloodwork and a complete diagnostic evaluation should be performed.

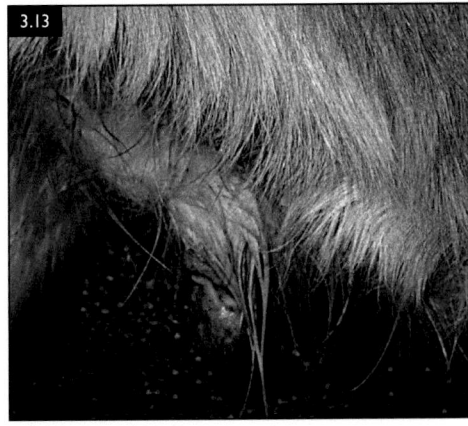

**FIG. 3.13** A foal with an acquired patent urachus showing a moist and swollen remnant of the umbilical cord through which urine is drained intermittently.

- ultrasonography may reveal:
  - umbilical remnant infection or dilated urachus continuous with the bladder.
- urinalysis and culture will identify urinary tract infection.
- contrast cystography has been used in more obscure cases.

## Management

- patent urachus cases, without concurrent umbilical disease, usually resolve spontaneously (often within a few days, but sometimes up to 14 days).
- prophylactic broad-spectrum antibiotics should be administered.
- application of desiccating or cauterising agents to the urachal opening should be avoided.
- concurrent or underlying disease should be treated.
- surgical removal of the umbilicus for uncomplicated patent urachus is rarely necessary.

## Umbilical hernia

### Definition/overview

- relatively common abnormality due to failure of the abdominal musculature to close around the umbilicus (congenital) or an acquired body wall defect.

## Aetiology/pathophysiology

- many normal foals at birth will have a defect in the body wall at the umbilicus:
  - palpable umbilical ring, which disappears over time.
  - not a congenital hernia until the foal is >1 month of age.
- acquired hernias arise between 5 and 8 weeks of age.
- may develop following omphalitis or omphalophlebitis.
- intestinal incarceration within a hernia in a foal, of whatever aetiology, is rare.
- larger hernial rings are less likely to close spontaneously:
  - greater chance of containing intestine or omentum.

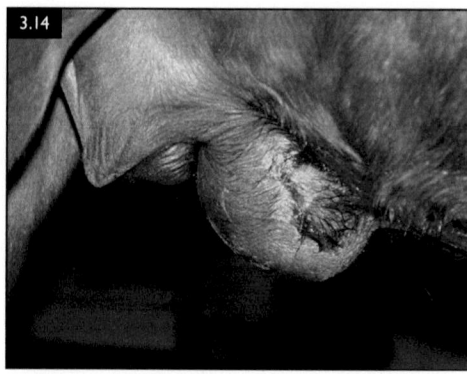

**FIG. 3.14** A young male foal with a congenital umbilical hernia. The hernial sac could easily be replaced into the abdomen and did not contain intestine at any stage.

## Clinical presentation

- visible and palpable, non-painful swelling of the umbilicus, which is reducible on palpation (Fig. 3.14).
  - muscular edge of the defect is readily palpable.
- time of detection depends on the age of defect development and the detail of foal observation but is often later, sometimes around weaning.
- no other clinical signs unless there is abscessation or loops of incarcerated bowel within the hernia (rare).

## Differential diagnosis

- umbilical infection or abscess
- ventral abdominal wall injuries.
- ventral oedema   • colic.

## Diagnosis

- palpation will determine:
  - size of hernia and its contents.
  - indication as to its reducibility.
  - rule out other umbilical disorders.
- ultrasonography can confirm the diagnosis in selected cases.

## Management

- small hernias may:
  - spontaneously shrink and close over time.
  - be carefully treated using elastrator bands or, in some countries, hernial clamps.

- this treatment is not favoured by all clinicians due to potential complications.
- can be a good option when surgery is not feasible.
- larger defects (6–8 cm) or those persisting at 6–9 months of age should be:
  - treated surgically, and a decision for castration may be made at this time.
  - very large hernias may require repair via insertion of a polypropylene mesh sub-peritoneally.

## Prognosis

- good for small hernias or those requiring simple surgical intervention.
- guarded for those that have incarcerated bowel or need mesh repairs.

## Uroperitoneum (See also **Book 4**)

## Definition/overview

- Uroperitoneum (urine in the peritoneal cavity) occurs in foals typically 1–5 days of age.
- Bladder rupture is the most common cause:
  - urachal, urethral, or ureteral defects may also be involved.
- clinical presentation depends on size, location, and aetiology of the defect.
- severe metabolic and electrolyte abnormalities are likely, and fatal if not corrected.

## Aetiology/pathophysiology

- pathophysiology in all clinical presentations of uroperitoneum is still not completely understood.
- proposed congenital aetiologies include:
  - developmental weakness of the dorsal bladder wall:
    - ♦ rupture may be spontaneous (possibly more likely in colts) and/or after improper/excessive handling of the foal's abdomen.
    - ♦ failure of the dorsal bladder wall to close during gestation.
    - ♦ ureteral ectopia with rupture, or ureteral/urethral atresia and rupture.
- acquired aetiologies include:
  - complication of neonatal intensive care due to:
    - ♦ prolonged recumbency or bladder distention.
    - ♦ septicaemia.
    - ♦ improper/excessive handling when standing or moving the foal.
    - ♦ focal necrotic cystitis/infectious urachitis secondary to ascending umbilical infections.
    - ♦ development of clinical signs in these cases may be insidious.
  - occasional foals with NMS cannot detect bladder distension and initiate a micturition reflex:
    - ♦ may require catheterisation for up to 7 days before function develops.
  - external trauma or strenuous exercise causing bladder avulsion from the urachus in older foals.

FIG. 3.15 A foal with uroperitoneum showing persistent posturing in order to urinate.

## Clinical presentation

- in normal foals, clinical signs develop at 2–3 days of age and can be non-specific.
- cases of infectious aetiology, clinical signs may be insidious and only apparent at 5–10 days of age.
- characteristic clinical signs include:
  - straining to urinate, dribbling urine, and frequent posturing to urinate (stretched out stance) (Fig. 3.15).
  - some foals may pass no urine or only minimal quantities (small defect).
- other clinical signs include:
  - lethargy, weakness, mild colic, and going off suck.

- progressive abdominal distension +/- palpable fluid thrill of the abdomen.
- tachypnoea and tachycardia:
  - ♦ later stages may be poor pulse quality, electrolyte-related dysrhythmias, and hypovolaemia.
  - ♦ with severe electrolyte disorders CNS signs, circulatory collapse, and death.
- occasional clinical signs include:
  - preputial or perineal oedema (particularly urethral and urachal ruptures).
  - urine accumulation in the scrotum.
  - protruding perineum.
- pyrexia is not a common feature unless infection is involved.

## Differential diagnosis

- meconium retention
- other causes of colic.

## Diagnosis

- history and careful physical examination, including abdominal palpation.
- urine from bladder catheterisation does not rule out uroperitoneum.
- serum biochemistry:
  - hyperkalaemia          hyponatraemia
  - hypochloraemia.
    - ♦ prior intravenous fluids may distort the classic electrolyte imbalances.
  - hypoglycaemia          acidaemia
  - azotaemia.
- transabdominal ultrasound:
  - large quantities of free, hypoechoic fluid are observed.

○ bladder is often small, irregularly shaped, and partially or completely collapsed.
○ defect in the bladder wall may be visible.
○ umbilical structures should also be examined.
- ultrasound-guided abdominocentesis to collect peritoneal fluid:
  ○ confirmed as urine:
    ♦ low cell count and specific gravity in uncomplicated cases:
      – without concurrent peritonitis or gastrointestinal compromise.
    ♦ peritoneal fluid creatinine concentration > twice that in peripheral blood.
- methylene blue in sterile solution can be infused into the bladder (via urethral catheter) and will be found in the peritoneal fluid if there is a defect in the bladder.
- retrograde contrast cystography and pyelography may be useful in diagnosing ureteral and urethral defects.

## Management

- uroperitoneum is initially a medical rather than surgical emergency.
- electrolyte abnormalities (particularly hyperkalaemia) put the foal at high anaesthetic risk.
  ○ **metabolic stabilisation is required prior to surgery.**
- urine should be gradually drained from the abdomen:
  ○ peritoneal drainage catheter can be left *in situ* until surgery.
  ○ alongside i/v fluid administration to support the circulation.
- acid–base and electrolyte abnormalities should be corrected with i/v fluid therapy:
  ○ 0.9% or 0.45% saline with 5% dextrose.
  ○ hyponatraemia should be corrected slowly.
  ○ **potassium-containing fluids should be avoided.**
  ○ hyperkalaemic foals:
    ♦ 5% dextrose saline is indicated (4–8 ml/kg/minute).
    ♦ severe and life-threatening hyperkalaemia (K+ >5.5 mmol/l [>5.5 mEq/l]):

– administer insulin (0.1–0.5 IU/kg s/c or i/v) or sodium bicarbonate (1–2 mEq/kg)
– moves potassium ions intracellularly and decrease serum potassium.
    ♦ calcium gluconate can be used as a transient cardioprotective.
    ♦ continuous ECG monitoring due to risk of arrhythmias and cardiac arrest.
- broad-spectrum antibiotics appropriate for use in foals are indicated.
- once foal metabolically stable:
  ○ surgical repair of the bladder should be performed.
  ○ umbilical remnants may be removed.
  ○ some foals are managed postoperatively with an indwelling urinary catheter for a short period of time.

## Prognosis

- good for uncomplicated bladder or urachal defects.
- second surgeries are occasionally required.
- poorer prognosis where concomitant infection or urinary tract defects are involved.

# Diarrhoea

## Definition/overview

- common clinical sign in foals.
- severity can vary:
  ○ transient self-limiting episode in a healthy foal.
  ○ life-threatening condition due to pathogen invasion or secondary complications.
- identification of the specific aetiology in individual foals can be challenging.
- treatment is often supportive in the physiological and non-infectious forms.

## Aetiology/pathophysiology

- colonic function in the young foal is immature:
  ○ more cases of diarrhoea are associated with small intestinal disorders.
- list of aetiologies is extensive:

## Physiological

- mild, self-limiting diarrhoea at 7–10 days old (also called 'foal heat diarrhoea').
- lasts for 1–7 days.
- associated with initiation of large intestinal microflora due to coprophagia and exploration of the foal's environment.
- diarrhoea is profuse and watery, but the foal remains well and on suck.

## Dietary

- milk overload
- inappropriate feeding of milk replacer.
- dietary indiscretion.
- lactose intolerance:
  - may be secondary to rotaviral or *Clostridium difficile* infections.

## Peripartum asphyxia

- foals with NMS from peripartum asphyxia:
  - may suffer gastrointestinal injury, leading to diarrhoea.
  - infectious causes are also possible in these cases.

## Bacterial

- *Clostridium difficile*, *Salmonella* spp., *C. perfringens*, *E. coli*, *Campylobacter* spp., *Rhodococcus equi*, *Bacteroides fragilis*, *Aeromonas hydrophilia*, and *Actinobacillus equuli* have been implicated.
  - *E.coli* produce enterotoxins affecting mucosal secretion.
  - *Salmonella* and *Clostridium* spp. are enteroinvasive and cause extensive mucosal damage.
    - loss of the mucosal barrier increases the risk of bacteraemia and septic shock.
    - impaired gut absorptive and secretory function.
  - intestinal spasm may lead to colic or ileus.
  - diarrhoea is often profuse and foetid.
  - Clostridial cases may have haemorrhagic faeces.
  - *Lawsonia intracellularis* is a cause of protein-losing enteropathy in older foals.

## Viral

- rotavirus and coronavirus affect the tips of the intestinal villi, leading to:
  - enzymatic maldigestion, malabsorption, and osmotic diarrhoea.
  - immature large intestinal function in the foal is unable to compensate.
  - rotavirus infections are common in suckling foals and highly contagious.

## Protozoal

- *Cryptosporidium parvum* most commonly causes gastroenteritis and diarrhoea in immunocompromised foals.

## Parasitic

- *Strongyloides westeri*, *Parascaris* spp., and large and small strongyles (more significant in the older foal and weanling).
- all can cause diarrhoea due to migrating larval damage, subsequent inflammation and vascular damage, motility disturbances, and protein leakage.
- colostral transfer of passive immunity is important in protecting foals against diarrhoea, particularly infectious causes and in the face of overwhelming pathogen challenge.
- pathophysiological consequences of diarrhoea in the foal can be extensive, with rapid and substantial deficits in fluid and electrolyte balance:
  - can quickly lead to circulatory shock, collapse, and death within hours.

## Clinical presentation

- full clinical history of the individual foal should be taken, to include age, feeding/sucking, nutrition, duration of signs, general demeanour, and serum IgG concentration.
- full disease epidemiology of the farm should also be known:
  - many infectious pathogens cause disease in outbreaks.
- detailed clinical examination should be performed including:
  - abdominal auscultation and assessment of hydration status.

- ○ diarrhoea itself (appearance, consistency, and volume).
  - ◆ haemorrhagic diarrhoea indicates an enteroinvasive pathogen.
- ○ abdominal pain can vary:
  - ◆ low grade with anorexia, to severe and mimics a surgical lesion.
  - ◆ signs of colic may precede the onset of diarrhoea.
- foals can quickly become dehydrated, hypovolaemic and have significant electrolyte imbalances.
- pyrexia, SIRS, sepsis, and circulatory collapse are rapid with some of the bacterial pathogens.
- renal dysfunction and gastroduodenal ulceration may occur as secondary complications.

## Diagnosis

- specific diagnosis often only possible in about 50% of cases:
  - ○ always attempted due to contagious nature of more common causes (e.g. rotavirus).
- fresh faecal samples placed in sterile containers can be used for:
  - ○ bacterial culture/sensitivity
  - ○ clostridial toxin ELISAs
  - ○ *Salmonella* PCR.
  - ○ viral qPCR/PCR/ELISAs
  - ○ *Cryptosporidium* ELISA.
  - ○ parasitological examination
  - ○ direct microscopy.
- blood samples are indicated in all foals with severe or chronic diarrhoea to include:
  - ○ haematology   ○ lactate   ○ glucose
  - ○ serum electrolytes.
    - ◆ leucopaenia indicates involvement of a bacterial pathogen +/- sepsis.
    - ◆ major electrolytes (sodium, potassium, chloride, calcium) may be reduced.
    - ◆ there may be a metabolic acidosis.
  - ○ serum proteins
  - ○ inflammatory markers
  - ○ renal parameters.
    - ◆ aid with diagnosis, assessment of disease severity, and directing therapy.
  - ○ volume and hydration status of the foal is determined using laboratory results alongside a detailed clinical examination, assessment of mentation, and evaluation of urine output.
- blood cultures should be taken in all young foals with diarrhoea and signs of SIRS or sepsis.
- repeated samples to indicate trends or responses are useful.

## Management

- **should be initiated without delay, especially for infectious cases.**
- treatment is similar in most cases and is generally supportive.
- detailed nursing plays a large role in successful outcomes.

### Fluid therapy

- **adequate and appropriate fluid therapy is key to successful treatment.**
  - ○ appropriate route and composition of fluids should be considered:
    - ◆ according to the severity of disease and metabolic disturbance:
      - – age of foal
      - – degree of mucosal damage suspected.
      - – severity of hypovolaemia and dehydration.
      - – metabolic disturbances
      - – milk intake.
    - ◆ compliance of the foal and competence of its handlers.
- oral fluids must be balanced electrolyte solutions plus glucose:
  - ○ achieved by adding commercially available products or calculated amounts of electrolytes to water.
  - ○ small volumes given frequently (up to 500 ml q1–2h) via stomach tube, dosing or bucket feeding.
  - ○ fresh clean water should always be available.
- intravenous fluids are important for rapid restoration of circulatory volume and correction of electrolyte imbalances:
  - ○ ideally administered by continuous infusion (requiring hospitalisation).
    - ◆ intermittent boluses can be given in practice.
  - ○ isotonic crystalloid should be used and spiked to make a 5% glucose solution:

- ♦ first litre as a resuscitation bolus in hypovolemic foals over 20 minutes.
- ♦ rate of infusion decreased to 3–5 ml/kg/hour accounting for maintenance requirements and increased losses.
  - o potassium can be added to the fluids but is safer as oral KCl (8 g/50 kg q8h) outside of a hospital setting.
  - o hyponatraemia should be corrected slowly.
  - o blood glucose concentrations should be monitored.
  - o i/v plasma administration should be considered to supplement:
    - ♦ albumin, passive immunity, metabolic nutrients, and clotting factors.
  - o foals with diarrhoea very rarely require bicarbonate administration:
    - ♦ metabolic acidosis is typically due to hyperlactataemia.
    - ♦ resolves with restoration of vascular volume and peripheral perfusion.

## Antiulcer medication

- prophylactic antiulcer medication in the foal remains controversial.

## Protectants/adsorbents

- many protectants/adsorbents are available for use in foals, including:
  - o di-tri-octahedral smectite (Bio-Sponge®; 3 tbsp in 30 ml of water p/o q6–12h).
  - o bismuth subsalicylate (Pepto Bismol; 1–2 ml/kg q8–12h).

## Drugs that reduce intestinal motility

- **contraindicated in infectious cases of diarrhoea.**
- chronic non-infectious cases, loperamide can be considered (0.1 mg/kg p/o q6h, then increased incrementally to 0.2 mg/kg).

## Probiotics

- many are available and may help re-establish gut microflora (scientific evidence is limited).
  - o increased incidence of diarrhoea in foals receiving a non-commercial probiotic containing *Lactobacillus pentosus* has been reported.

## Analgesics and anti-inflammatory drugs

- Flunixin meglumine has been used for analgesia and to treat endotoxemia (0.25 mg/kg i/v q8–12h)
  - o high risk of side effects, especially in collapsed and/or hypovolaemic foals.
- Butorphanol (0.01–0.04 mg/kg i/v) is a useful analgesic:
  - o short acting and has sedative effects.
- N-butylscopalammonium bromide (Buscopan™) (0.3mg/kg i/v) is also used.

## Antibiotics

- **antibiotics are generally contraindicated except in:**
  - o young foals with sepsis and diarrhoea.
  - o those at risk of sepsis.
  - o cases of salmonellosis.
- appropriate first line combinations are noted in the 'Neonatal septicaemia' section (See page 178).
  - o based on culture and sensitivity results where available.
  - o knowledge of the pharmacokinetics of the drugs used is essential:
    - ♦ excretion/secretion into the gut and effect on gut flora.
  - o Metronidazole (10–15 mg/kg p/o q6–12h) is indicated for clostridial diarrhoea and early use positively affects outcome.
  - o toxic side effects of antibiotics are potentiated in the hypovolaemic foal.

## Exogenous lactase enzyme

- given to foals to aid in digestion with:
  - o rotavirus, coronavirus, and *Cryptosporidium*.
  - o primary lactose intolerance or osmotic diarrhoea.
  - o administered at 6000–9000 IU/50 kg foal q3–8h.

## Feeding

- milk is typically not withheld unless:
  - o severe cases of diarrhoea (+/- ileus or colic)
    - ♦ 24 hours of dietary rest may be required.
    - ♦ administer intravenous glucose and monitor blood glucose concentrations.
- parenteral nutrition may be necessary in severe cases, but requires hospitalisation.

- bolus stomach tubing or, ideally, the use of an indwelling feeding tube will allow for administration of enteral nutrition, when the foal is off suck.
  - starting volume of 300–400 ml for a Thoroughbred foal under 2 weeks of age.

## Nursing

- high standards of nursing and hygiene are important:
  - foals should be kept clean, dry, and comfortable.
  - regular cleaning of soiled areas and applications of barrier ointments.
  - **advise handlers of potential zoonoses (Salmonella, Campylobacter) and the highly contagious nature of some infectious diarrhoeas.**

## Other drugs

- anticoagulants (low molecular weight heparin) may be beneficial in sepsis cases.
- Polymixin B can be used as an anti-endotoxic drug (potential for nephrotoxicity).
  - usually reserved for hospitalised cases.

## Control and prevention

- biosecurity measures include:
  - protective clothing, gloves, and foot dips.
  - keeping each mare and foal in their designated stable throughout treatment.
  - stabling or using dedicated contaminated 'dirty' paddocks for scouring foals helps reduce environmental build-up of pathogens.
- rotavirus vaccine for pregnant mares is available that increases passive immunity in the foal but is not fully protective against disease.

## Septic arthritis/osteomyelitis

## Definition/overview

- common, potentially career- or life-threatening condition seen in young foals:
  - typically, within the first 30 days of life.
  - part of other multifocal infections.
- early diagnosis and appropriate aggressive treatment are key to a successful outcome.

## Aetiology/pathophysiology

- infection via the haematogenous route is most common in foals:
  - due to bacteria seeding from a focal infection (e.g., omphalitis).
  - from the digestive/respiratory systems.
  - secondary complication of septicaemia.
- foals with neonatal disease or adverse peripartum events are at the highest risk.
- dissemination of pathogenic bacteria in the joint itself or nearby bony epiphysis/physis during transient or persistent bacteraemia:
  - facilitated by increased and slow blood flow with low oxygen tensions to both rapidly growing bones and synovium.
  - may also spread from osseous to soft tissue, and vice versa.
- traumatic wounds and iatrogenic sources of infection are uncommon.
- multiple joints and/or physes/epiphyses may be involved.
- most commonly isolated bacteria are similar to those recorded in septicaemic foals.
- older foals, *Rhodococcus equi* may also lead to:
  - septic epiphysis/physis and/or, to an immune-mediated non-septic synovitis.
- established infection causes marked exudative septic inflammation leading to rapid ischaemic necrosis, bone and/or cartilage degeneration, then soft tissue fibrosis and abscess formation in the bone.

## Clinical presentation

- 5 different disease presentations are recognised according to the anatomical site affected, the aetiology of the initial infection, the subsequent clinical signs, radiographic changes, and necropsy findings.
  - may not be always easy to clinically differentiate them in individual cases.
- S type:
  - septic synovitis in one or more swollen/painful joints without bone involvement.
  - acute-onset severe lameness in foals less than 14 days old.
  - may be systemic illness and minimal radiographic signs evident.

- **E type:**
  - infection in one or more joints and adjacent epiphysis.
  - slightly older foals (3–4 weeks), with acute-onset severe lameness.
  - stifle and tarsocrural joints are most affected:
    - ◆ joint distension, considerable periarticular swelling, and deep bone pain.
  - radiographic abnormalities are often evident.
  - prognosis is more guarded.
- **P type** (Fig. 3.16, 3.17):
  - infection occurs in the physis and metaphysis, usually at a single site.
    - ◆ adjacent joint may become infected or develop a sympathetic non-septic effusion.
    - ◆ less common than the S type.
  - foals from 1–12 weeks of age
  - may be a prior history of systemic illness.
  - swelling and pain on palpation over the physeal region:
    - ◆ with or without joint distension.
  - repeated radiographs may be necessary to show abnormalities in some cases.
  - distal physis of the long bones are most affected.

- **I type:**
  - infection enters the joint from infected periarticular soft tissues (abscesses).
  - upper limb joints are most affected (i.e. the coxofemoral and femorotibial joints).
  - early detection of soft tissue infection helps prevent this serious complication.
- **T type:**
  - rare cases involving infection of small cuboidal bones of the tarsus and carpus:
    - ◆ collapse of these bones allows spread of infection to the joint or joints.
    - ◆ **separate condition to aseptic cuboidal bone disease in premature foals.**
    - ◆ joint swelling, moderate to severe lameness, and radiographic changes.
- some affected joints are not always easily palpable:
  - shoulder or hip joint.
  - all joints should be carefully examined.
- foci of infection should be identified, and systemic or multifocal disease investigated.

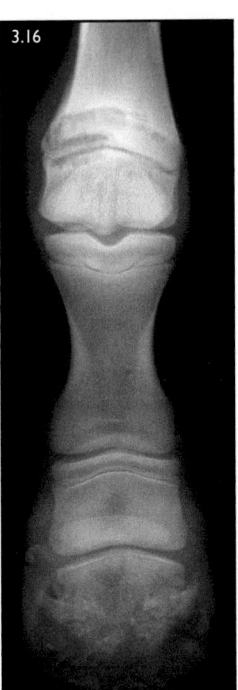

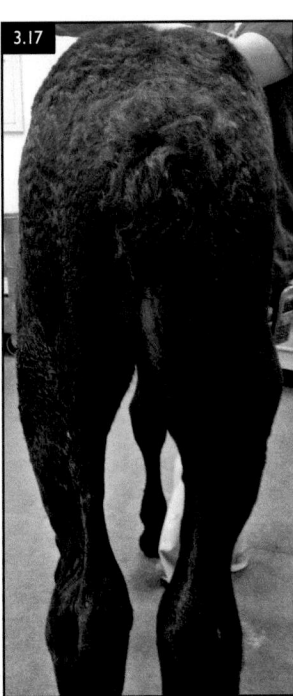

**FIG. 3.16** Dorsopalmar radiograph of the distal limb of a 4-week-old foal with septic physitis of the lateral distal third metatarsal bone (P type).

**FIG. 3.17** 8-week-old foal with P type osteomyelitis affecting the right stifle with obvious swelling on the lateral aspect.

## Differential diagnosis

- other causes of infectious and non-infectious lameness:
  - fractures       foot abscess
  - traumatic injuries.
  - non-septic joint disease
  - haemarthrosis.
- young foals presenting with two or more relevant clinical signs, should be treated as a potential infection until proven otherwise.

## Diagnosis

- confirmed by analysis of joint fluid obtained by aseptic synoviocentesis:
  - visual examination, total and differential WBC counts, and total protein analysis.
- synovial fluid and blood culture and sensitivity testing are important:
  - false-negative results are common.
- material retrieved from bone lesions should be cultured and examined cytologically.
- joint and physeal region radiography helps differentiate:
  - type of disease, chronicity, direct therapy, and clarify prognosis.
  - early cases may have false-negative findings:
    - repeat radiography should be considered 3–10 days later.
- joint and physeal region ultrasonography can be useful especially in periarticular swelling.
  - umbilical remnants should be examined when investigating for infection foci.
- blood samples will usually show an increased WBC count and acute phase proteins.

## Management

- early treatment is essential due to the rapid and progressive degenerative changes in an infected joint and the potential life-threatening nature of disease.
- considerable economic implications in treating these cases and this should be discussed.
- **Antibiotics:**
  - immediate broad-spectrum, bactericidal systemic antibiotics with good distribution in joints and bone at suitable dose rates without delay.
  - culture and sensitivity results can guide subsequent choices if necessary.
  - continue administration until at least a week after full disease resolution.
  - toxicity of drugs, including aminoglycoside nephrotoxicity should be considered.
  - intra-articular antibiotics can be useful:
    - may cause increased joint inflammation.
    - post lavage (i.e. 150–300 mg gentamicin or amikacin) can used once or repeated.
    - antibiotic-impregnated collagen sponges can be left in the joint to provide sustained drug release.
    - intra-articular catheters for continuous antibiotic infusion may be considered for concurrent epiphyseal osteomyelitis.
  - intravenous regional limb antibiotic perfusion and/or intraosseous medication are useful in non-responsive cases.
- **Lavage:**
  - joint lavage is essential to remove infectious and inflammatory material:
    - through-and-through needle lavage technique in the sedated or anaesthetised foal using 3–5 litres of sterile polyionic fluid per joint (Fig. 3.18).
    - surgical lavage may be more appropriate.
    - different techniques have associated advantages, disadvantages, and costs.
- **Surgery:**
  - joint lavage via arthroscopy is the most thorough and effective treatment:
    - allows removal of abnormal material/synovium (can be cultured).
    - opportunity to debride cartilage and bone lesions.
    - establishment of postoperative drainage.
    - assessment of prognosis.
  - poorly responding cases, arthrotomy and open drainage can be effective.
    - postoperative management is expensive and prolonged.

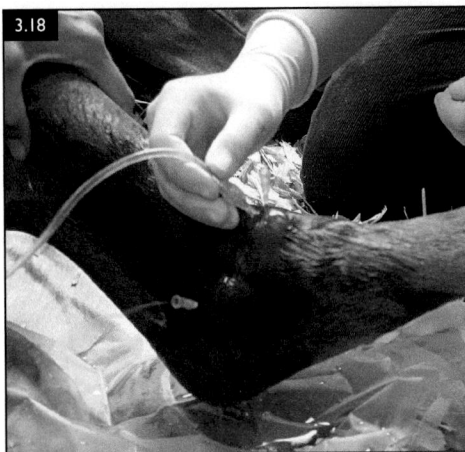

**FIG. 3.18** The tarsocrural joint of a young Thoroughbred foal with a septic synovitis under-going through-and-through lavage of the joint under heavy sedation and local regional analgesia.

- surgical curettage of septic physes followed by open drainage and lavage:
  - ◆ can be successful but is expensive and may lead to pathological fractures.
- **Other treatments:**
  - concurrent systemic problems in the foal should be diligently addressed.
  - provision of supportive nursing care and ongoing monitoring.
  - judicious use of NSAIDs may be warranted, especially in systemically ill foals.
  - foal should be box rested and affected limbs bandaged.
  - repeat lavage in septic arthritis may be necessary.
  - osteomyelitis – serial measurement of SAA useful in monitoring treatment response:
    - ◆ reduced concentrations can precede improvement in the lameness.

## Prognosis

- dependent on the number of sites of infection, the degree of damage to the joint(s), the presence of osteomyelitis, concurrent systemic disease and the response to therapy:
  - involvement of multiple joints and/or bones adversely affects the outcome.
  - presence of systemic illness similarly decreases the prognosis.

- chronic cases and those poorly responsive to initial aggressive treatment often require extensive and prolonged treatment.
    - ◆ poor prognosis for a return to full soundness.
- acute S type cases treated quickly and aggressively carry the best prognosis.

# Respiratory disease in the older foal

## Definition/overview

- bronchopneumonia is a significant cause of morbidity and mortality in foals aged 1–8 months.
- clinical signs are often mild until the disease process is advanced:
  - mild cases may resolve spontaneously.
- outbreaks of infectious respiratory disease are relatively common.

## Aetiology/pathophysiology

- bacterial causes are the most common. Pathogens include:
  - *Streptococcus zooepidemicus*
  - *Streptococcus equi.*
  - *Rhodococcus equi*   ○ *Pasturella* spp.
  - *Bordetella bronchiseptica.*
  - *Pneumocystis jiroveci*   ○ *E. coli*
  - *Actinobacillus* spp.
- viral agents are frequently associated with secondary bacterial disease:
  - include EHV-1 or EHV-4, equine influenza virus, equine adenovirus, and equine rhinovirus.
- other less common causes in individuals include:
  - aspiration pneumonia due to cleft palate or dysphagia.
  - congenital cardiac abnormalities.
  - traumatic conditions such as fractured ribs or diaphragmatic hernia.
  - parasite infestations such as hepatopulmonary migration of ascarid larvae.
  - atypical interstitial pneumonia.

## Clinical presentation

- clinical signs vary in range and severity but include:
  - tachypnoea (resting rate >25 breaths/minute).

- ○ abnormal respiratory pattern and effort.
- ○ cough (often heard when the foal is being turned out):
  - ♦ amount of coughing does not correlate well with degree of disease.
- ○ nasal discharge:
  - ♦ typically, not in *Rhodococcus equi* bronchopneumonia.
- ○ pyrexia   ○ depression   ○ lethargy.
- ○ anorexia and poor growth rate.
- ○ progression to respiratory distress.
- foals may not exhibit clinical signs until disease is advanced.
- pulmonary auscultation does not correlate well with the severity of disease.
- extrapulmonary manifestations of *R. equi* infection include:
  - ○ intra-abdominal abscessation, uveitis, non-septic polysynovitis, and others.

## Diagnosis

- knowledge of the premises respiratory disease epidemiology is useful.
- thorough physical examination will identify pertinent clinical signs.
- clinical pathology:
  - ○ haematology and inflammatory markers will indicate infection:
    - ♦ leucocytosis   ♦ neutrophilia.
    - ♦ increased SAA and fibrinogen concentrations.
    - ♦ chronic cases may have anaemia.
  - ○ tracheal aspirates are stressful for foals:
    - ♦ cytology and culture of samples helpful for specific diagnoses and targeted antimicrobial therapy in some cases.
      - – false-negative culture is quite common.
    - ♦ sample cytology and VapA gene PCR provides definitive diagnosis of *R. equi* bronchopneumonia.
- thoracic ultrasonography is easy to perform and useful to identify (Fig. 3.19):
  - ○ parenchymal disease extending to the lung surface
  - ○ pulmonary abscesses.
  - ○ pleural effusions, pleuritis, and haemothorax.

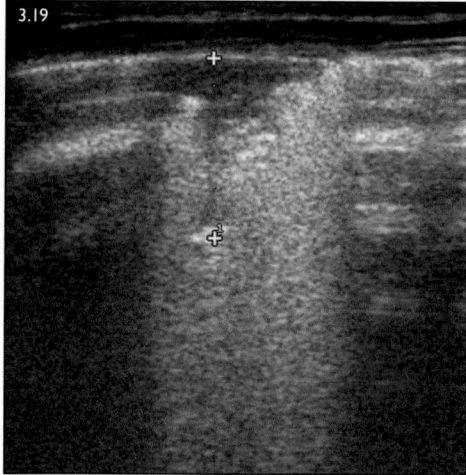

**FIG. 3.19** Trans-thoracic ultrasonography of the thorax of a foal with aspiration pneumonia showing focal consolidated parenchyma.

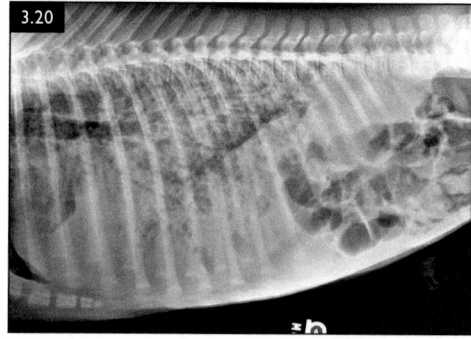

**FIG. 3.20** Laterolateral radiograph of a foal with severe abscessing bronchopneumonia.

- thoracic radiography can additionally assess the degree and type of disease, its severity and progression (Fig. 3.20).
- endoscopy of the upper respiratory tract – possible with care and scopes <9 mm diameter.
- arterial blood gas analysis may be appropriate in severe cases.
- full post-mortem examination should be carried out on any foal that dies.

## Management

- **Antibiotics:**
  - ○ broad-spectrum antibiotics from the outset:

- culture and sensitivity results can guide subsequent choices if necessary.
- research continues regarding the optimal treatment for *Rhodococcus equi* bronchopneumonia:
  - current recommendations remain to use the macrolides alongside rifampicin:
  - clarithromycin/rifampicin combination is the most effective.
  - azithromycin is useful for foals that are difficult to treat and benefit from a prolonged inter dosing interval.
  - macrolides are associated with severe or fatal colitis from weanling age onwards and this should be discussed.
  - current dose rates are:
    - Clarithromycin: 7.5mg/kg p/o q12h.
    - Azithromycin: 10 mg/kg p/o q24h.
    - Rifampicin (never use as monotherapy): 5–10 mg/kg p/o q12h.
  - other causes of bacterial pneumonia, clarithromycin can be used alone, or the following antimicrobials may all be used:
    - Trimethoprim sulfa: 15–30 mg/kg p/o q12h.
    - Ceftiofur: 5–10 mg/kg i/m or i/v q12h.
    - Cefquinome: 1 mg/kg i/m or i/v q12h.
    - Oxytetracycline: 5–10 mg/kg i/v q12h.

- Doxycycline: 10 mg/kg p/o q12h.
  - Trimethoprim sulfa (30 mg/kg p/o q12h) is the only antibiotic effective against *Pneumocystis jiroveci* and should be used until complete disease resolution.
- **Increasing clearance of secretions:**
  - bronchodilators may be helpful in some foals.
- **Environmental management:**
  - avoid excessive handling/stress.
  - provide a cool, airy environment.
  - reduce levels of dust and respiratory irritants.
  - avoid hyperthermia:
    - risk in high environmental temperatures for some foals receiving macrolides.
  - avoid transport unless essential.
- **Respiratory support:**
  - severely compromised foals may benefit from humidified intranasal oxygen therapy.
- **Anti-inflammatory drugs:**
  - NSAIDs may be used judiciously at low doses and for short periods in severe cases.
  - parenteral or nebulised corticosteroids may be beneficial in interstitial pneumonia.

## Prognosis

- good for uncomplicated bacterial and *R. equi* bronchopneumonia:
  - most survivors race as well as age-matched peers.
- guarded for most cases of atypical interstitial pneumonia.

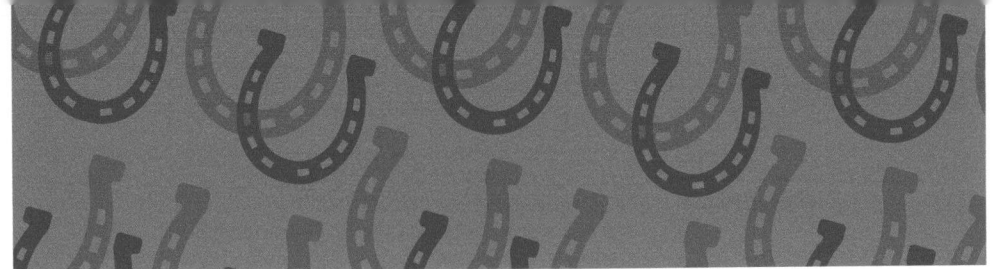

# Index